WOLLASTON

People Resisting Genocide

WOLLASTON
People Resisting Genocide

MILES GOLDSTICK
Foreword by Dr. Rosalie Bertell

BLACK
ROSE
BOOKS

Montréal • New York

Black Rose Books No. P 111

Canadian Cataloguing in Publication Data

Goldstick, Miles, 1955-
Wollaston: people resisting genocide

Includes index.
ISBN 0-920057-94-2 (bound). —ISBN 0-920057-95-0 (pbk.).

1. Uranium mines and mining—Social aspects—Saskatchewan—Wollaston Lake
Region. 2. Uranium mines and mining—Environmental aspects—Saskatchewan—
Wollaston Lake Region. I. Title.

HD9539.U72W34 1987   304.2'8   C87-090040-4

Cover design: Jennifer DeFreitas
Layout and pasteup: Kai Mikalsen
Photo screens: Mattias Lentz and Lars Jacobsen

The author has donated all royalties (about $1 Canadian per copy) to the anti-
nuclear/native rights movement.

Black Rose Books

3981 boul. St-Laurent   340 Nagel Dr.
Montréal, Qué. H2W 1Y5   Cheektowaga, N.Y. 14225
Canada   USA

Printed and bound in Québec, Canada

Acknowledgements

This book would never have been completed without the help and encouragement of many people. I first want to thank Margareta Åkerström, Frederik Leidegren, and Melanie Monsma for sharing a home with me that sometimes became more of an office. Thank you also to Stephanie Sydiaha for being such a hardy northern traveller and inexhaustible source of energy, despite living in Saskatoon – the administrative centre of the Saskatchewan uranium industry.

Layout artist Kai Mikalsen worked with patience and perseverance. Evald Bjerkli and Runar Forseth assisted with the layout. Dag Viljen Poleszynski donated office space and accommodation for the final layout work. Nature and Youth in Oslo, Norway contributed some layout and reprographic materials. *Lillebror* (Little Brother) The Centre For Critical Documentation And Action, in Oslo, kindly loaned their light table. A small amount of financial support was received from The Peoples' Movement Against Nuclear Power and Nuclear Weapons in Sweden, and Rita A. Burtch of Smiths, Ontario. SoftCraft donated the computer software to print the second draft.

Many amateur photographers made important contributions. My apologies for not including the photo credits with every photo. Professional photgraphers Ritva Kovalainen and Raimo Long from Finland, and Rune Eraker and Helge Hummelvol from Norway, donated their photos. An extra effort was made in photo production work by professionals Ross Brown, Lars Hillborg, and Gerrie Warner. Others helping with photo production include Helen Leidegren, Assar Lindberg, Michael Manley-Casimir, Karin Qvarnström, and Karen Rose of *The Peak* (Simon

Fraser University student newspaper). Kjerstin Andersson helped with the graphics work for the figure of uranium mill wastes produced to make fuel used by a nuclear reactor.

Maisie Shiell had many helpful comments on "The Mines." Diana Leis and Alar Olljum influenced the section titled "Saskatchewan, The Saudi Arabia Of The Uranium Industry." Olle Holmstrand had useful comments on "What About The Wastes." Adele Ratt assisted with the "The Chipewyan People" section. Stephanie Coe helped document the chronology of events during the June, 1985 gathering and blockade.

Most of the statements by northern people were transcribed from tapes recorded by Ritva Kovalainen, Kia Lundqvist, Nancy McPherson and April Page. Transcriptions were done by Stephanie Coe, Dan Kay, April Page, Adele Ratt, Jeanne Shaw and Stephanie Sydiaha.

Of invaluable assistance were comments by Thijs De la Court, Simon Dalby, Wilma Groenan, Marianne Grödum, Jim Harding, Kia Lundqvist, Beverly McBride, Henrik Persson, Frannie Ruvinsky, Jan Stoody, Stephanie Sydiaha, Katarina Yale, Joanne Young and Jan Öhman. I am especially grateful to Alar Olljum for helping to edit the final manuscript. Thank you also to the handful of people who contributed significantly and asked not to be acknowledged, some out of fear of losing their jobs.

Most deserving of recognition are all the participants in and supporters of the June 1985 gathering and blockade at Wollaston Lake. They have helped to keep the spirit of resistance alive.

Finally, I must add that responsibility for any errors and mistakes are mine alone.

To
David Garrick,
master environmental activist living in Vancouver, B.C.,
who works unceasingly in defense of the Earth,
despite sometimes ''going uphill through molasses,''
as he puts it.

Table Of Contents

Foreword

By Rosalie Bertell, Ph.D., G.N.S.H.

There are many similarities between a community's response to a health problem and an individual's response. How many times have we waited to see if a pain would "go away," put off seeing a doctor because we were too busy, or fudged a little on a health form to get a more lucrative job? Clearly, ill health is a social and economic burden to the individual. It is also burdensome for a community to say: "our land and food is polluted; we are experiencing increases in cancers, birth defects and other illnesses." We all need friends to help us to face harsh realities as well as groundless fears in such real life problems.

Suppose you are a dairy farmer and you fear that your dairy cattle are producing inferior and/or polluted milk. Your vet bills soar; your herd suffers from infertility, still-births and deformed offspring. As you try to cope, it appears to you to be suicidal to launch a campaign against the industry polluting your grazing land because if you do the market for your milk, your only source of income, will plummet, whether it is polluted or not. It will be an economic disaster.

A situation intermediate between the individual and the community health problem might help to elucidate the new problems. Suppose a restaurant is found by the Board of Health to be responsible for ptomaine poisoning. The results of their failure in good food handling are dramatic and immediate. The management is temporarily embarrassed, but the situation is salvageable and good food management can be initiated. Healthfulness is restorable. In the interim, the public has access to other sources of food and other restaurants. A short term solution to the problem is available.

After the Chernobyl accident, April 26, 1986, however, many European governments ordered dairy cows to be kept off pasture because of radiation contamination. By late April the winter store of hay in the barns had been exhausted. One by one, farmers applied for and got permission to pasture their cows outside rather than force them into malnutrition and death. What should then be done with the contaminated milk? In England it was dried and powdered, so that it could be held in storage until the iodine-131 had decayed. The other radioactive iodines, the strontium-90, cesium-137 and some ten other radionuclides remained in the powdered milk. In Sweden, contaminated milk was poured over the farmland, its radioactive contamination allowed to reconcentrate in germinating new crops. Some countries mixed contaminated milk with uncontaminated milk, "adulterating" the latter and providing a more uniformly bad product to the growing child and pregnant mother. These practices were "justified" with elaborate risk analyses, matching trade-offs between the dollar value of "illness" and the dollar value of substituting uncontaminated milk and of compensating farmers for losses not due to their mismanagement. "Permissible" levels of pollution were defined by governments.

At Wollaston Lake a health problem may be posed by moose grazing near radioactive tailings piles from the uranium mines. The moose provide an important source of meat for the native people and are a tourist attraction for American and Canadian recreational hunters. Eating contaminated moose does not bring sudden dramatic illnesses, rather it slowly undermines health, like old age does, and causes the children in a community to be a little less healthy than their parents. Slowly over time, the vigor of the people is irreversibly reduced.

In situations where food is contaminated, should the people who will be forced to bear the brunt of the health, economic and social penalties be made to decide to seek the information? Does the restaurant owner call the Board of Health or the farmer invite the radiological investigative team to examine his herd's milk?

What is the federal and provincial responsibility to the native people and to the tourists to assure the healthfulness of the food chain? What is the responsibility of the uranium mining companies? What good are treaties safeguarding the right to hunt and fish if the game and fish are not safeguarded?

Food is the basis of human life. Starvation, botulism, ptomaine poisoning and toxic contamination of food cause death at various speeds and through different mechanisms. Death however, is still death. We as a society have failed the Wollaston Lake community in the basic human right to information and to health. We are multiplying comparable self-destructive behaviors all over the world. Please read on because the story of Wollaston Lake is our own story, the story of our land, our food, and our brothers and sisters.

Introduction

The uranium industry is more active in northern Saskatchewan than any other place in the western world. For the native people of the area it is the dominant force continuing the destructive momentum built up over 300 years of colonialism. If the present trend continues, the result will be genocide. Mining and exploration are taking place though native land claims and aboriginal rights have not been settled.

Wollaston Lake is one of the small Indian communities struggling to survive. In the summer of 1985 the people there hosted an unprecedented protest. From June 14-17, on the west side of Wollaston Lake, all traffic in and out of the Rabbit Lake and Collin's Bay uranium mines was blocked for 80 hours. The blockade marked the first coordinated act of civil disobedience against the uranium industry in Saskatchewan.

The decision to carry out the blockade was made only after all conventional means of protest had been exhausted. More than a decade of public demonstrations, letter writing campaigns and court cases all failed. Despite repeated invitations, industry representatives refused to appear in any open public debate with the opposition. Instead, mine expansion and industry public relations activities accelerated. The corporate/government strategy is clear: hush up all the problems of uranium mining and keep the issue out of the media.

The blockade was significant for several reasons. First, it was a response to the appeal of the indigenous people of Wollaston Lake, who initiated and acted as spokespeople during the action. Second, it marked the first coordinated mobilization. Approximately 200 native people participated in the blockade, coming from villages in northern Saskatchewan and Manitoba. About 35 non-native activists came from across Canada. Support protests took place in Toronto; Montreal, Copenhagen, Denmark and Dublin, Ireland, as support mail flowed in from all over the world.

Furthermore, this was the first act of resistance to uranium mining in Saskatchewan to receive widespread media attention, spurring corporate/government officials to respond with both coercive and co-optive means. False rumors and accusations about the position of Wollaston people abounded. The most common accusation was that the resistance to uranium mining was the result of manipulation by outside agitators. By providing only a glimpse of northern peoples' own statements and the problems of uranium mining, this book proves that the real "outside agitators" are the mining companies. The documentation is not meant to be comprehensive. Lack of resources have ensured that.

The first two chapters, "The People" and "The Mines," contrast two very different relationships to the Earth: dominance and destruction of the land by the uranium mines versus Indian people trying to live in harmony with their surroundings.

"The People" chapter has four main parts. First is a short description of the community of Wollaston Lake and the people there. The ancestors of the Wollaston Chipewyan people signed Treaty Number 10 with the Crown of England in 1906. A copy of the Treaty is reprinted. Some information about the church and Father Megret, the priest in Wollaston, is included. The second and third parts are, respectively, statements by Mayor George Smith of Pinehouse and Brian "Banjo" Ratt from Ile à la Crosse. Their statements on colonialism and neo-colonialism come from direct experience as they live in communities closest to the world's largest uranium mine, Key Lake. The final part of "The People" is a statement by a government employee in the prairie provinces (Alberta, Saskatchewan, and Manitoba) who has extensive experience helping Indian people in trouble with the law.

In "The Mines" an overview of mining activity in northern Saskatchewan is given, along with the waste problem it has created. Some brief comments are made on the closed, operating and planned mines. Included as well is an explanation of some connections between uranium and nuclear weapons; and information about uranium bullets.

The waste problem is looked at in detail in "What About The Wastes." The Rabbit Lake/Collin's Bay operation is used as a focus. The records of the other two operating mines in Saskatchewan, Key Lake and Cluff Lake, are certainly worse due to their greater size and higher grade ore. Six major topics are covered, including: the nature of the problem, effects on plants and animals, no solution to stopping contamination, possible remedial action, and the illegal operation of the Rabbit Lake mine. The last part is a review of the position of the authorities and Eldorado Nuclear Limited regarding pollution from uranium mining. Their response to the growing controversy around the Collin's Bay mine is included. It is shown that they often try to hide the serious pollution problems caused by uranium mining. They clearly state the pollution is "not significant."

13

Finally, a summary of the conclusions of this chapter are given.

Chapter 3, "The Resistance," is made up mostly of testimonies by Wollaston residents and other people about uranium mining, rather than in-depth literature research as in "The Mines." The story of the June 1985 gathering and blockade is given after a summary of the history of opposition to uranium mining by Wollaston residents from 1972 to April 1985. Reprinted is the submission by the Wollaston Local Advisory Council (LAC) to the La Ronge, July 1981 public hearing on the Collin's Bay B-zone project.

Following then are the minutes of a community meeting in Wollaston on April 30, 1985 with representatives of Saskatchewan churches. One purpose of the church leaders going to Wollaston was to prepare for a meeting with the Saskatchewan government. The church leaders requested that Saskatchewan Premier Grant Devine sit down and talk with them and northern spokespeople. Devine refused.

The part on the June 1985 gathering and blockade begins with a short description of the Collin's Bay Action Group, the coordinating group. Then come some resolutions and letters of support against uranium mining written before the blockade took place. The events of June 11–17 are documented in a day by day chronology. Some of the support activities are noted that took place during that time in Saskatoon, Toronto, Ottawa, and Dublin, Ireland. Included is a statement made by Bill Blaikie, MP (New Democratic Party) in the House of Commons debates June 17. A portion of a journal kept by Gerry Paschen, one of the blockaders, is reprinted.

After the daily chronology, the outcome and a short critique of the blockade is given. In "The Outcome" is a transcription from part of a tape recording made at a press conference held in Saskatoon on June 20th by native leaders. Sol Sanderson, the pro-uranium mining president of the Federation of Saskatchewan Indian Nations (FSIN), dominated the press conference. Reprinted then is a short statement against uranium mining and supporting the blockade made the next day, June 21, by Jim Manly, MP (New Democratic Party), in the House of Commons debates. Included as well is a letter opposing uranium mining from the Saskatchewan Association of Northern Local Governments (SANLG). Ending "The Outcome" is a statement by anti-uranium mining advocate George Smith, Mayor of Pinehouse and Chairman of SANLG.

"A Critique" covers only a few aspects of the gathering and blockade. An essay by Jack Ross discusses the question of non-

14

violence at the blockade as compared to Gandhi's principles.

The last chapter summarizes some international solidarity during August and September 1985. Most of the chapter is statements by members of Scandinavians Against Nuclear Development (SAND). They are a group of 11 Scandinavians who made a three week tour of northern Saskatchewan. Sweden and Finland purchase uranium from Saskatchewan mines. Also mentioned are Adele Ratt's trip to Japan and the Norwegian anti-uranium multimedia production "Iktoms Profeti."

Adele Ratt is from the northern Saskatchewan uranium boom town of La Ronge, and is a spokesperson for the Saskatchewan native rights and anti-uranium movements. She travelled to Japan to participate in the 40th anniversary commemoration of the Hiroshima and Nagasaki bombings.

The book ends with an afterword by Adele Ratt, titled "Rationale and Proposal for a Northern Development Council."

Chronology of Events
July 1984—August 1985

July 13, 1984: Open letter written by Wollaston Lake Lac La Hache Band and Local Advisory Council appealing for support to stop uranium mining.

December 1984: Collin's Bay Action Group (CBAG), a coalition of native rights and anti-uranium groups, formed for the purpose of coordinating the June 1985 gathering and blockade at the entrance to the Rabbit Lake and Collin's Bay uranium mines.

March 26, 1985: The first of three Eldorado public relations mine tours for residents of Wollaston Lake. The other two tours took place on April 3 and 11, 1985.

April 17, 1985: Environment Canada and Atomic Energy Control Board (AECB) representatives participated for the first time in a community meeting in Wollaston Lake, though mining had been going on for a decade across the lake.

April 30, 1985: Community meeting in Wollaston Lake with representatives of the Inter-church Uranium Committee (ICUC). A letter was sent to Premier Grant Devine from the ICUC asking for a meeting to discuss the Wollaston situation. Devine refused.

May 6, 1985: Marie Rose Yooya as representative of the Lac La Hache Band expressed the Band's anti-uranium mining sentiments to the Federal Minister of Environment in Ottawa; and participated in a workshop with anti-nuclear groups from across Canada.

May 27, 1985: A court case against Eldorado Nuclear for breaking the conditions of their Rabbit Lake mine operating license, issued by the AECB, was "stayed" in La Ronge provincial court. The provincial Attorney General's office ordered the case not be heard. A "stay" completely bypasses the judge and the government is under no obligation to give any reason for its action.

June 1, 1985: A couple kilometres south of the entry to the Rabbit Lake and Collin's Bay mines Eldorado Nuclear erected a new gate across the road, and a fence extending about 40 metres on

16

both sides.

June 3, 1985: Delegates to the annual meeting of the United Church of Canada picketed the Eldorado Nuclear office in Saskatoon.

June 11 and 12, 1985: Meetings in Wollaston between outside supporters and community members filled the Band Hall to capacity.

June 13, 1985: Community meetings concluded. About 150 people moved from the community across the lake to camp at the Umperville River campground, just south of the mine gates.

June 14–17, 1985: About 150 people blocked the road to the Rabbit Lake and Collin's Bay uranium mines at the entrance gate.

June 20, 1985: In Saskatoon a meeting was held between northern native leaders and corporate/government officials. A number of Dene Chiefs from northern Saskatchewan attended, as did representatives of Eldorado and the federal and provincial governments. The northern leaders held a press conference after the meeting. It was announced that a commitment was made between the Lac La Hache Band and Eldorado to form an "Economic Committee" that would determine how the Wollaston people would benefit from mining.

June 27, 1985: Mine manager Michael Babcock announced to the media that a three hour meeting at the mine with Chief Kkailther and the Band Councillors was "very successful," and that plans were being made to solve their communication problem.

July 10, 1985: Les Erikson of SMDC met with Councillors of the Lac La Hache Band and the Wollaston LAC to inform them of SMDC's summer exploration activities in the Wollaston region. After the short meeting in the Band Hall the group went on a tour of the Cigar Lake mine site.

August 1, 1985: A group of 11 people from Scandinavia arrived in Saskatoon to see first hand the effects of uranium mining on the people and land. Their three week tour included the communities of Ile à la Crosse, Pinehouse, La Ronge, Southend and Wollaston Lake.

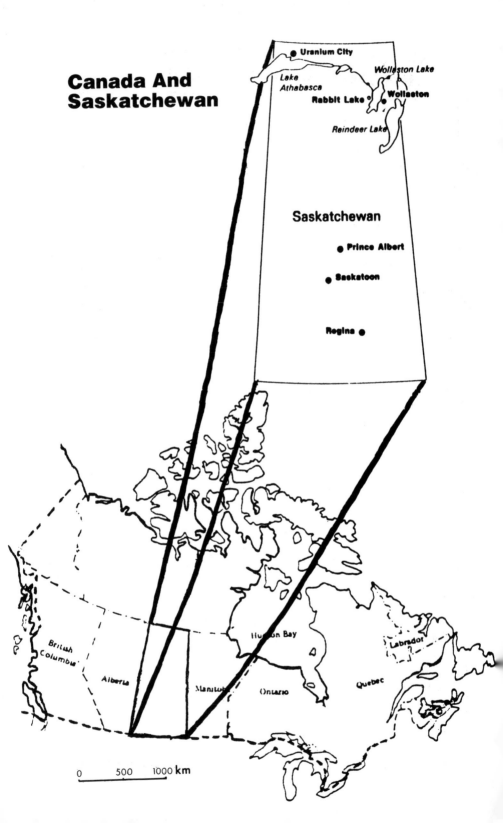

Canada And Saskatchewan

● Uranium City

Lake Athabasca

Wolleston Lake

Rabbit Lake ○   ● **Wollaston**

Reindeer Lake

Saskatchewan

● **Prince Albert**

● **Saskatoon**

Regina ●

British Columbia

Alberta

Manitoba

Ontario

Hudson Bay

Quebec

Labrador

0   500   1000 **km**

Uranium Mining Activity In Northern Saskatchewan

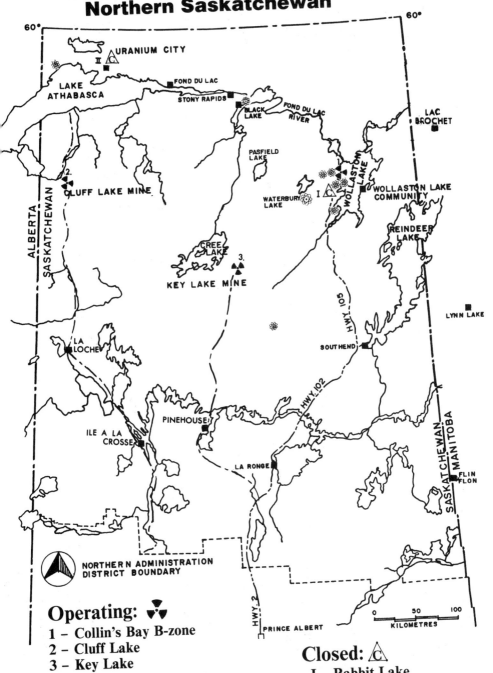

Operating: ☢
1 – Collin's Bay B-zone
2 – Cluff Lake
3 – Key Lake

Under Construction: ⚛

Closed: Ⓒ
I – Rabbit Lake
II – Beaverlodge And Others

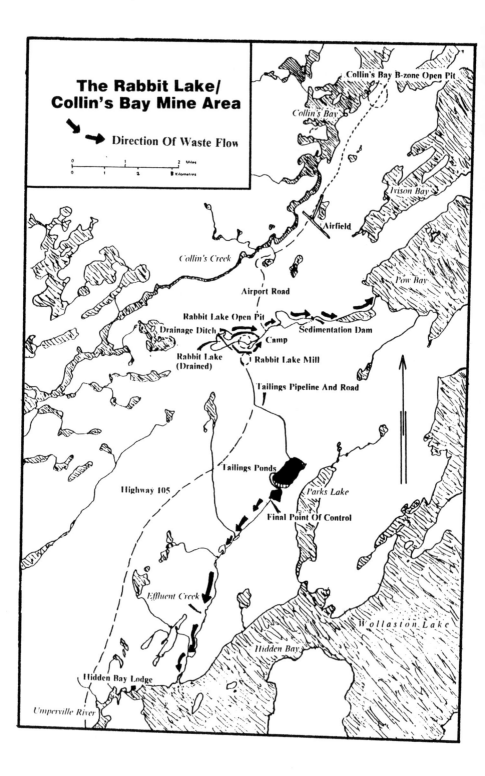

The Rabbit Lake/
Collin's Bay Mine Area

Direction Of Waste Flow

Collin's Bay B-zone Open Pit

Collin's Bay

Ivison Bay

Airfield

Collin's Creek

Pow Bay

Airport Road

Rabbit Lake Open Pit

Drainage Ditch

Sedimentation Dam

Camp

Rabbit Lake (Drained)

Rabbit Lake Mill

Tailings Pipeline And Road

Tailings Ponds

Parks Lake

Highway 105

Final Point Of Control

Effluent Creek

Wollaston Lake

Hidden Bay

Hidden Bay Lodge

Umperville River

CHAPTER 1
THE PEOPLE

Introduction

About 30,000 people live in northern Saskatchewan, an area 300,000 square kilometres in size. Approximately 20,000 are of native ancestry. They live in 35 villages and small towns.[1] While billions of dollars have been spent on the uranium industry, the basic needs of Indian communities are not being met. The statistics tell the story:

- In some northern communities close to 90% of the people are unemployed. Less than 1/3 of the overall working population have jobs. The situation is critical as about half the native people in the north are under the age of 20.
- Though native people make up less than 10% of the total population, they represent more than 60% of the people in Saskatchewan prisons.
- Less than 5% of all the people in Canada are dependent on social assistance but more than 75% of indigenous people must rely on it.
- A 1980 inspection by Environment Canada of sewage treatment facilities on Saskatchewan Indian reserves found that about 75% did not comply with federal guidelines.[2]
- Healthcare and education are drastically inadequate. Alcoholism has reached an epidemic level.

Diana Ralph, Ph.D., assistant professor at the Faculty of Social Work, University of Regina prepared a report giving an overview of health conditions in northern Saskatchewan. The 1984 report documents that in the north as compared to the whole province:

- there are over twice as many people per doctor;
- payments for physician services per patient are less than one third as much;
- life expectancy is about 10 years less;

The barge landing where highway 109 passes the west side of Wollaston Lake.

 - the death rate from all causes is almost 50% higher;
 - infant deaths are about 20% higher;
 - cancer deaths are about 30% higher; and further, the report states:

Northern Treaty Indians are hospitalized 61% more often than the average Saskatchewan resident. Since 1975, hospitalizations for cancer, birth defects and circulatory illnesses have increased dramatically (between 123 and 600% in the northern population aged 15 to 64 - the entire labor force). At the same time, there is a large increase in hospitalizations among young children for digestive disorders and birth anomalies.[3]

The rest of this chapter contains some information about the community of Wollaston Lake, statements by Pinehouse Mayor George Smith and Brian "Banjo" Ratt from Ile a la Crosse; and an interview by a government employee in the prairie region who talks about the difficulties Indian people have in the cities.

Wollaston Lake: A Portrait

Wollaston Lake is an isolated Indian community of about 800 Chipewyan and Metis people situated along the shoreline of Welcome Bay on the southeastern edge of Wollaston Lake. The community is located at about 58 degrees north latitude and 103 degrees west longitude. It is known that there was a village southwest of the present town in the 1940's. In about 1947 the village suffered a smallpox epidemic and the survivors fled.

Today's town site was chosen in 1957 because it was thought at that time that a road would be built there from the south. Instead, the Saskatchewan government built highway 109 to facilitate the transnational exploitation of uranium deposits on the other side of the lake, 40 km to the west. The road was completed in 1974. Almost every road in northern Saskatchewan has been built to extract resources, primarily uranium, as is typical of colonial development. Other transportation links to Wollaston include scheduled stops by Norcanair, a commercial airline company, and a summer barge service between the community and "barge-landing," the point where highway 109 is closest to the west side of the lake.

By 1970 the Wollaston community had a population of about 300, which doubled before 1980. In 1985 there were about 10 old age pensioners in the town and about two-thirds of the population was under 16 years old. People born before the mid-1950's do not read, write, or speak English well. Men are more literate than women.

The lake provides the indigenous people with their drinking water and forms the basis of the local economy through the traditional pursuits of fishing, hunting and trapping. Most people eat fish from the lake year-round. The lake is truly the heart and life blood of the community.

In terms of commercial fishing, the lake is one of the most important in northern Saskatchewan and is given a one million pound per year quota. It is the largest single waterbody wholly within the borders of Saskatchewan.

There are few wage jobs for the native population, and groceries and gasoline are expensive. The meagre income from commercial fishing and the local supply of food and firewood are

The barge from the community to the west side of the lake.

the mainstay of the community.

In 1979 the provincial government introduced a subsidy program covering part of the costs of flying in perishable goods to remote northern communities. The program was cancelled on January 15, 1985. The next day prices went up about 30%.[4]

There is a very short growing season in Wollaston, though small summer gardens have had some success. Snow can stay on the ground and ice on the Lake until the end of June. Then after a short summer the snow returns by October.

Wollaston Lake is comprised of approximately 75% status Chipewyan Indians of the Lac La Hache Band who are included under Treaty Number 10. A "status" Indian is a person registered as Indian under the terms of the Indian Act. The criteria for registration are historical and legal rather than racial.[5] Status Indians are governed by the federal Ministry of Indian Affairs. The remaining 25% are non-status Chipewyans from the Brochet area and Métis of Cree descent from the Pinehouse region. The Métis people moved gradually north during the late 1950's in response to government programs to develop the commercial fishing potential of Wollaston Lake. The non-status and Métis

26

A main road in Wollaston.

people are under provincial government authority and elect a
council that has limited local powers called a Local Advisory
Council (LAC).

The division in the community between provincial and federal government jurisdiction is the source of many problems. The
two levels of government have different housing, healthcare,
education and other policies. People with the same ancestry and
culture are often treated very differently.

Wollaston is the only permanent settlement in the extreme
northeastern portion of Saskatchewan. The closest large town is
Lynn Lake, Manitoba, 190 kilometres by air to the southeast.
The significance of Lynn Lake to Wollaston waned after the
completion of highway 109. Since then the southern centre most·
used by Wollaston residents became La Ronge (population 2,579
– October 1, 1986).

It wasn't until the early 1970's that the community got electricity (by large diesel generators) and June 1977 that telephone
arrived. The CBC North TV station became available in 1980. In
1984 a satellite dish was installed by the LAC giving one channel
to a part of the community, and in 1985 the LAC began oper-

27

ation of a local FM radio station for a few hours per day. The radio reception from outside the community is very poor.

Poverty

The impact of colonialism on the Wollaston people has been profound. A consulting group hired by the federal Ministry of Indian Affairs carried out a planning study in 1981. The summary of their report reads:

> The old adage about "hewers of wood and drawers of water" is a fact of daily life in Wollaston Lake community. The settlement is years behind other northern Saskatchewan communities in the provision of basic services and facilities and it is decades behind other, comparable, southern Saskatchewan communities. What makes this statement surprising and perhaps shocking, is that the community is only 25 years old and, as such, is one of the youngest communities in the whole of the Province. *By most standards of community prosperity, this community can count itself amongst the neglected few.*
>
> ...The most serious infrastructure deficiency found in the study was the extremely poor quality of existing housing and the severe overcrowding which is experienced by local families. Average occupancy levels are more than double the Saskatchewan normal rate. (Emphasis added.)

As is typical of northern communities, a couple of dozen southerners live in Wollaston. They are the priest, teachers, nurses, businessmen, pilots, Royal Canadian Mounted Police (RCMP) and government wildlife resource agents. The RCMP moved into Wollaston in 1984. None of the indigenous people's homes are connected to running water and indoor sewage. With the exception of the priest Jean Megret, all of the southerners' homes have indoor plumbing. However, the school, medical clinic, and LAC Hall do have running water and indoor sewage.

A further expression of poverty in the community is that the lack of proper sewage and garbage disposal has resulted in a serious hepatitis threat every spring. A hepatitis epidemic occurred in the spring of 1984. The problem is sewage runoff and seepage from a misplaced garbage dump contaminating Welcome Bay, the people's drinking water supply. The natural turnover of water in the narrow bay is too slow to adequately dilute the waste runoff that rushes in when the snow melts.

The LAC office and post office.

The RCMP office, complete with small jail cell.

The Lac La Hache Band office.

29

Commercial fishermen taking their catch to barge landing.

The Chipewyan People

The Wollaston Chipewyan people are descendants of the Caribou Eater branch of the Dene Nation. They were named Chipewyan, meaning "pointed hood" in Cree, by Cree people to the south because of the pointed hoods they made on their garments.

The Chipewyan language is rich and difficult to master. Margaret Reynolds, author of "The Dene Language Book," writes,

> The Dene (Chipewyan) language is the third hardest language in the world to speak and understand. It has 39 consonants and 116 vowel sounds. The language is guttered and nasalized with tongue tip trills.[6]

There is no word in Dene for uranium. In 1977, a Dene elder who had learned of some of the dangers of uranium coined a phrase for it: "dada-thay," meaning "death rock."

As recently as 1950, and still fresh in the memory of the Elders, the Chipewyan people had a nomadic lifestyle following the migrating Beverly Caribou herd. The Chipewyan survived in

30

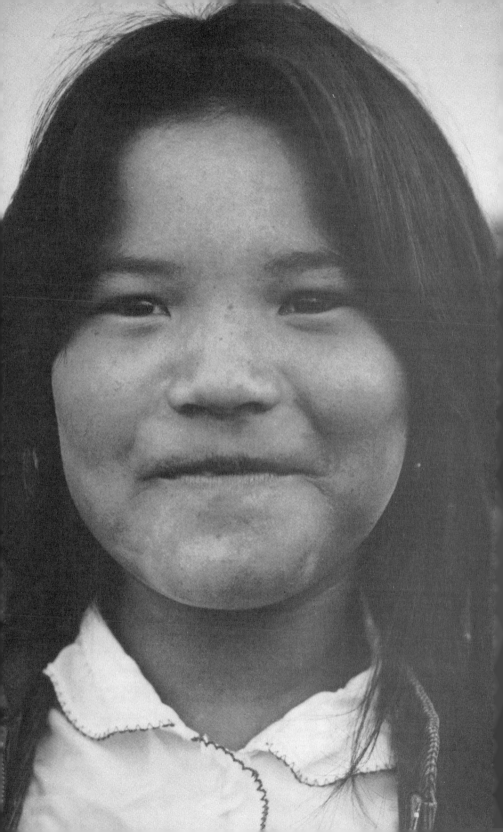

Cutting up caribou meat.

harmony with the land by hunting, fishing and gathering. Cultural and spiritual traditions were passed on orally from generation to generation in legends, song and dance. Hand drums were a part of every social activity and are a means of communication with the spirits and all of the natural world.

Many aspects of modern life have served to draw the people away from their rich cultural heritage. Such as: the education system, dependency on government services, family breakdown due to alcoholism and drug abuse, and the influence of the Roman Catholic church (which says that the traditional spiritual ways are little more than witchcraft). Many of the traditions were lost during the British colonial days of the fur trade and treaty making by the British government. The colonizing process accelerated in the late 1940's when the government established a fur trading post and fish packing plant on Moose Island, just west of the present community. The colonizers encouraged the local people to rely on imported food and supplies, and government sponsored "aid." The Wollaston people, like other indigenous peoples, had been self-sufficient for thousands of years.

The Lac La Hache Band and non-status and Métis people were confined on a reservation and settlement in the same manner as other tribes and have come to rely heavily on government

33

Late September 1985: there is already snow on the ground.

facilities. However, the Wollaston people still maintain their life-style of hunting, fishing and gathering. The tradition of the drum is still alive, though it is slowly being lost as the elders pass on.

The strong cultural ties to the land are evident in the continued use of small settlements in the Wollaston area, such as Jack Pine Island near the mouth of the Cochrane River at the north end of Wollaston Lake, an area which previously marked the southern limit of the Chipewyan migration route. Each spring entire families migrate north to these settlements – a symbol of their independence and of the enduring strength of traditional ways.

Kinship ties with other Chipewyan communities are maintained despite the great distances involved. These communities include Black Lake and Stony Rapids to the northwest, Kinoo-sao to the southeast, Cree Lake to the southwest and the Brochet area of Manitoba to the east.

Joanne Young, from her summer, 1984 visit to Wollaston Lake:

I have read a great deal about pornography and its harmful effects but until my stay at Wollaston Lake I had no personal experience with it. Some questions and conversation with children there brought a rude awakening.

Apparently a wide variety of "hard porn" video cassettes have found their way into some of the homes and are watched by the children. Never having been far from this tiny isolated village, they are unfamiliar with life in the rest of Canada. The children seem to assume that these films depict "normal" customs in the big towns and cities, and so they are prepared to behave similarly themselves!

Hard pornography is strongly influencing the standards of behavior and expectations these children form for their future lives. Isn't it bad enough that we have destroyed the native culture of these people, have herded them together into villages where they are bored to death, are "developing" nearby lands in such a way as to make their future lives in this area impossible, without imposing our twisted moral and spiritual values on their children?

35

Treaty Number 10

Treaty Number 10 was signed between the Lac La Hache Band and the Crown of England in 1906.[7] At a community meeting in Wollaston on April 30, 1985, Wollaston Elder Bart Dzeylion stated,

My mother is still alive. She is 97 years old. She was alive when the treaty was signed. I keep care of her now. She cannot walk too good now, but she wanted me to say she would like to be here. I told her I'm going to meetings and hearing how the land is being destroyed. She responded,
I was there when the treaty was signed and there were discussions for a week and we were told, "We're coming to you to sign the treaty, not to harm you in any way. By signing this treaty we are agreeing to help and take care of you people. We are in no way going to harm the way you live."
Why are they doing this now? We made agreements. If they are going to do these things to us why did they get us to sign the treaty in the first place?

At the same meeting Wollaston Chief Hector Kkailther said,

From my personal observations, hearing people and studying the treaty agreements, it seems as if Indian Affairs had a field day.

37

They did whatever they wanted to do. When the treaty was signed there wasn't anyone who could write or read English. It looks like the government did this to trick us. We have to study and understand about the treaties and how they affect us.

The text of the treaty provides for the setting aside of "reserves of land." However it was not until 1970 that the present Lac La Hache Reserve No. 220 was established, though it was surveyed in 1965.[8] There was no community participation program to determine the reserve boundaries, which split the community into two sections of land – one under provincial and the other under federal jurisdiction. Several treaty Indian families discovered that their homes were off the reserve when they received provincial government tax notices.

The "Wollaston Lake Community Planning Study" done in 1981 reads in the summary:

According to the Band's interpretation of the treaty it retains sovereign powers over reserve territory and it maintains jurisdiction over a range of other lands in the northeasterly region of Saskatchewan. Such extraterritorial rights include hunting, trapping and gathering areas, fishing stations, burial grounds, sacred lands, meeting grounds and timber births. Despite the wording of the treaty the Band contends there was no consent reached on a variety of natural resources namely minerals, game, forests, water and even airspace. As such, the Bank contends that without their consent, through negotiation and agreement, the activities in the area including mineral exploration, mining, tourist outfitting, road construction and crown land leasing in general are in violation of their treaty rights.

The whole treaty signing process is a clear example of past, immoral and unethical manipulation by the colonizers. This was

not an isolated incident. Just as extreme oppressive treatment has continued to the present day. The entire treaty is reprinted below. Note that each Indian has an "X" registered on the document as their signature.

There is no question that the validity of Treaty Number 10 could be challenged legally in a number of different ways. The fact that none of the Indians could speak, read or write English when the treaty was signed is one example. However, in practical terms it is extremely difficult for a community having few high school graduates, let alone any lawyers, to take on the Canadian judicial system.

TREATY No. 10

Articles of a treaty made and concluded at the several dates mentioned therein, in the year of our Lord one thousand nine hundred and six between His Most Gracious Majesty the King of Great Britain and Ireland by His commissioner, James Andrew Joseph McKenna, of the city of Winnipeg, in the province of Manitoba, Esquire, of the one part, and the Chipewyan, Cree and other Indian inhabitants of the territory within the limits hereinafter defined and described by their chiefs and headmen hereunto subscribed of the other part.

Whereas the Indians inhabiting the territory hereinafter defined have, pursuant to notice given by His Majesty's said commissioner in the year 1906, been convened to meet His Majesty's said commissioner representing His Majesty's government of the Dominion of Canada at certain places in the said territory in this present year 1906 to deliberate upon certain matters of interest to His Most Gracious Majesty on the one part and the said Indians of the other.

And whereas the said Indians have been notified and informed by His Majesty's said commissioner that it is His Majesty's desire to open for settlement, immigration, trade, travel, mining, lumbering and such other purposes as to His Majesty may seem meet, a tract of country bounded and described as hereinafter mentioned and to obtain the consent thereto of his Indian subjects inhabiting the said tract and to make a treaty and arrange with them so that there may be peace and good will between them and His Majesty's other subjects, and that His Indian people may know and be assured of what allowances they are to count upon and receive from His Majesty's bounty and benevolence.

And whereas the Indians of the said tract, duly convened in council at the respective points named hereunder and being requested by His Majesty's said commissioner to name certain chiefs and headmen who should be authorized on their behalf to conduct such negotiations and sign any treaty to be founded thereon and to become responsible to His Majesty for the faithful performance by their respective bands of such obligations as shall be assumed by them, the said Indians have therefore acknowledged for that purpose the several chiefs and headmen who have subscribed hereto.

And whereas the said commissioner has proceeded to negotiate a treaty with the Chipewyan, Cree and other Indians inhabiting the said territory hereinafter defined and described and the same has been agreed upon and concluded by the respective bands at the dates mentioned hereunder;

Now therefore the said Indians do hereby cede, release, surrender and yield up to the government of the Dominion of Canada for His Majesty the King and His successors for ever all their rights, titles and privileges whatsoever to the lands included within the following limits, that is to say:—

All that territory situated partly in the province of Saskatchewan and partly in the province of Alberta, and lying to the east of Treaty Eight and to the north of Treaties Five, Six and the addition to Treaty Six, containing approximately an area of eighty-five thousand eight hundred (85,800) square miles and which may be described as follows:—

Commencing at the point where the northern boundary of Treaty Five intersects the eastern boundary of the province of Saskatchewan; thence northerly along the said eastern boundary four hundred and ten miles, more or less, to the sixtieth parallel of latitude and northern boundary of the said province of Saskatchewan; thence west along the said parallel one hundred and thirty miles, more or less, to the eastern boundary of Treaty Eight; thence southerly and westerly following the said eastern boundary of Treaty Eight to its intersection with the northern boundary of Treaty Six; thence easterly along the said northern boundary of Treaty Six to its intersection with the western boundary of the addition to Treaty Six; thence northerly along the said western boundary to the northern boundary of the said addition; thence easterly along the said northern boundary to the eastern boundary of the said addition; thence southerly along the said eastern boundary to its intersection with the northern boundary of Treaty Six; thence easterly along the said northern boundary and the northern boundary of Treaty Five to the point of commencement.

And also all their rights, titles and privileges whatsoever as Indians to all and any other lands wherever situated in the provinces of Saskatchewan and Alberta and the Northwest Territories or any other portion of the Dominion of Canada.

To have and to hold the same to His Majesty the King and His successors for ever.

And His Majesty the King hereby agrees with the said Indians that they shall have the right to pursue their usual vocations of hunting, trapping and fishing throughout the territory surrendered as heretofore described, subject to such regulations as may from time to time be made by the government of the country acting under the authority of His Majesty and saving and excepting such tracts as may be required or as may be taken up from time to time for settlement, mining, lumbering, trading or other purposes.

And His Majesty the King hereby agrees and undertakes to set aside reserves of land for such bands as desire the same, such reserves not to exceed in all one square mile for each family of five for such number of families as may elect to reside upon reserves, or in that proportion for larger or smaller families; and for such Indian families or individual Indians as prefer to live apart from band reserves His Majesty undertakes to provide land in severalty to the extent of one hundred and sixty (160) acres for each Indian, the land not to be alienable by the Indian for whom it is set aside in severalty without the consent of the Governor General in Council of Canada, the selection of such reserves and land in severalty to be made in the manner following, namely, the Superintendent General of Indian Affairs shall depute and send a suitable person to determine and set apart such reserves and lands, after consulting with the Indians concerned as to the locality which may be found suitable and open for selection.

Provided, however, that His Majesty reserves the right to deal with any settlers within the bounds of any lands reserved for any band or bands as He may see fit; and also that the aforesaid reserves of land, or any interest therein, may be sold or otherwise disposed of by His Majesty's government of Canada for the use and benefit of the Indians entitled thereto, with their consent first had and obtained.

It is further agreed between His Majesty and His said Indian subjects that such portions of the reserves and lands above mentioned as may at any time be required for public works, buildings, railways or roads of whatsoever nature may be appropriated for such purposes by His Majesty's government of Canada due compensation being made to the Indians for the value of any improvements thereon, and an equivalent in land, money or other consideration for the area so appropriated.

And with a view to showing the satisfaction of His Majesty with the behaviour and good conduct of His Indians and in extinguishment of all their past claims, He hereby through His commissioner agrees to make each chief a present of thirty-two (32) dollars in cash, to each headman twenty-two (22) dollars and to every other Indian of whatever age of the families represented at the time and place of payment twelve (12) dollars.

His Majesty also agrees that next year and annually thereafter for ever He will cause to be paid to the Indians in cash, at suitable places and dates of which the said Indians shall be duly notified, to each chief twenty-five (25) dollars, each headman fifteen (15) dollars and to every other Indian of whatever age five (5) dollars.

Further His Majesty agrees that each chief, after signing the treaty, shall receive a silver medal and a suitable flag, and next year and every third year thereafter each chief shall receive a suitable suit of clothing, and that after signing the treaty each headman shall receive a bronze medal and next year and every third year thereafter a suitable suit of clothing.

Further His Majesty agrees to make such provision as may from time to time be deemed advisable for the education of the Indian children.

Further His Majesty agrees to furnish such assistance as may be found necessary or advisable to aid and assist the Indians in agriculture or stock-raising or other work and to make such a distribution of twine and ammunition to them annually as is usually made to Indians similarly situated.

And the undersigned Chipewyan, Cree and other Indian chiefs and headmen on their own behalf and on behalf of all the Indians whom they represent do hereby solemnly promise and engage to strictly observe this treaty in all and every respect and to behave and conduct themselves as good and loyal subjects of His Majesty the King.

They promise and engage that they will in all respects obey and abide by the law; that they will maintain peace between each other and between their tribes and other tribes of Indians and between themselves and other of His Majesty's subjects whether whites, Indians, half-breeds or others now inhabiting or who may hereafter inhabit any part of the territory hereby ceded and herein described, and that they will not molest the person or trespass upon the property or interfere with the rights of any inhabitant of such ceded tract or of any other district or country or interfere with or trouble any person passing or travelling through the said tract or any part thereof and that they will assist the officers of His Majesty in bringing to justice and punishment any Indian offending against the stipulations of this treaty or infringing the law in force in the country so ceded.

In witness whereof His Majesty's said commissioner and the chiefs and headmen have hereunto set their hands at Isle à la Crosse this twenty-eighth day of August in the year herein first above written.

Signed by the parties hereto in the presence of the undersigned witnesses the same having first been explained to the Indians by Magloire Maurice, interpreter.

J. V. Begin,
    Supt., R.N.W.M. Police.
I. Rapet, ptre, O.M.I.,
Chas. Fisher,
Chas. Mair,
Angus McKay,
D. McKenna,
T. Davis.

J. A. J. McKENNA,
Commissioner.
his
WILLIAM X APISIS,
mark
Chief of the English River Band.
his
JOSEPH X GUN,
mark
Headman.
his
JEAN BAPTISTE X ESTRAL-SHENEN, mark
Headman.
his
RAPHAEL X BEDSHIDEKKGE,
mark
Chief of Clear Lake Band.

Signed by the Chief and Headman of
the Canoe Lake Band this 26th day of
September, A.D. 1906. The treaty
having been read over and explained
by Archie Park, interpreter, in the
presence of the undersigned witnesses.
J. V. BEGIN,
    Supt., R.N.W.M.P.,
L. COCHIN, ptre, O.M.I.,
J. E. TESTON, ptre, O.M.I.,
F. E. SHERWOOD,
    Const., R.N.W.M. Police,
     his
ARCHIE X PARK, Interpreter.
  mark
CHARLES MAIR,

           his
    JOHN X IRON,
     mark
Chief of Canoe Lake Band.
        his
  BAPTISTE X IRON,
     mark
Headman, Canoe Lake Band.

          his
JEROME X COUILLONEUR,
      mark
Headman, Canoe Lake Band.

Articles of a treaty made and concluded at the several dates mentioned therein, in the year of our Lord one thousand nine hundred and seven, between His Most Gracious Majesty the King of Great Britain and Ireland by His Commissioner Thomas Alexander Borthwick, of Mistawasis, in the province of Saskatchewan, Esquire, of the one part, and the Chipewyan, Cree and other Indian inhabitants of the territory within the limits hereinafter defined and described by their chiefs and headmen hereunto subscribed of the other part.

     *   *   *   *   *   *   *

In witness whereof His Majesty's said commissioner and the chiefs and headmen have hereunto set their hands at Lac du Brochet this 19th day of August, in the year first above written.

Signed by the parties hereto in the presence of the undersigned witnesses the same having first been explained to the Indians by A. Turquetil.
    CHARLES LA VIOLETTE,
         Interpreter.
W. J. McLEAN, *Witness.*
A. W. BELL, *Witness.*
THOMAS BORTHWICK,
    Commissioner, Treaty No. 10.

          his
  PETIT X CASIMIR,
      mark
Chief of Barren Land Band.
         his
  JEAN X BAPTISTE,
      mark
Headman of Barren Land Band.
         his
ANDRE X ANTSANEN,
      mark
Indian of Barren Land Band.

In witness whereof His Majesty's said commissioner and the chiefs and headmen have hereunto set their hands at Lac du Brochet this 22nd day of August in the year first above written.

Signed by the parties hereto in the presence of the undersigned witnesses the same having first been explained to the Indians by E. S. Turquetil, interpreter.
  Witness A. W. BELL,
    "   W. J. McLEAN.

          his
THOMAS X BENAOUNI,
      mark
Chief of Hatchet Lake Band.
  Witness A. W. BELL,
         his
  PIERRE X AZE.
      mark
Headman of Hatchet Lake Band.

     THOS. BORTHWICK,
       Commissioner, Treaty 10.

44

A structure for smoking meat and fish, with wind protection.

The Wollaston church.

The Church and Father Megret

The impact of the church on the north should not be under-estimated. In the Americas, as on other continents, church representatives prepared the way, culturally, ideologically and economically, for colonialism. Missionaries were crucial in the process of disenfranchisement, notably through their role as translators during the signing of treaties between the Chipewyan and the British government.

The church in northern Saskatchewan does not practice liberation theology as is taking place in Latin America. It is important to point out, however, that outside of the north the church is split on the uranium mining issue and development policy in general. Many church leaders in southern Saskatchewan, including Prince Albert, have spoken out strongly against uranium mining.

A church service.

Photo: Ritva Kovalainen

In Wollaston Lake most of the people are devoutly Roman Catholic. They often attend church twice a day and take part in an annual pilgrimage to Lac St. Anne, Alberta, just west of Edmonton. Jean Megret, the local priest, is clearly the modern personification of the colonial missionary. Megret lobbies strongly and openly in support of uranium mining, and is quick to defend the nuclear industry in his home country, France.

Father Megret visits the Rabbit Lake mine site regularly and keeps in touch with the French management of the Cigar Lake Mining Corporation. He does not hide the fact that in the late 1970's he received an all expense paid trip to Hawaii from Gulf Minerals (then owner of the Rabbit Lake mine). In the months before the June 1985 gathering and blockade, mining companies subsidized several jaunts to Saskatoon for him to meet with company executives.

Needless to say, Megret was an outspoken opponent of the

gathering and blockade. In fact, he left town in early June and did not return until late June, threatening to never come back "unless all the commotion stopped."

The following interview of Father Megret was made by Scandinavians 'Against Nuclear Development (SAND) when they visited Wollaston Lake in mid-August, 1985:

Father Megret:

I arrived in this country, in the north, in 1947. After the war. I went first to La Loche and Dillion, close to Ile à la Crosse and after that I moved to this area, but my home was Brochet, east of Wollaston. I got established here in the 1950's. The first place was at the north end of the lake about 40 miles from Wollaston. Then we moved to Wollaston in the late 1950's.

The first church was built in 1957, we got the other one in 1973. At that time the people were all nomads. They were scattered 200 miles from here. In the spring they got together and their base was Brochet. They were following the caribou, living practically strictly on trapping and hunting and a little bit of fishing when there was no meat around. It wasn't until the mid 1960's that people lived in Wollaston year-round.

People are living at the mine site. They don't seem to be more sick than people in Wollaston. It would be much better if they would hire half the people in Wollaston to go and work at the mine. Then there would be more employment. But there are two problems. The lack of a minimum of education, at least being able to speak and read English; and also the willingness. I know for a fact that some don't want to work.

Personally, it's my own opinion, I know there is a lot of people who don't agree with me but I don't think there is any direct danger. One thing that I know is that they are very careful and they have inspectors from federal and provincial governments on their back all the time. You just have to trust the people in charge. But another thing also, you can't stop them. You cannot stop those uranium mines. They'll take it where they find it and nobody can stop them. Anyway, I don't see how it would be better to stop them.

What will they do with the plutonium? That's not my problem.

48

Father Jean Megret, the priest in Wollaston. "You just have to trust the people in charge."

49

Statement By Mayor George Smith, Pinehouse[9]

We always supported ourselves and when the mines go, we'll be supporting ourselves. We've never gone hungry yet. This is the best place in the world. This is God's land. Nobody can starve here, but the mining makes it more difficult.

If you look at Wollaston Lake, there's trappers there that have lost their traplines. Pretty soon more traplines will be taken out. Wherever you destroy the land somebody is going to lose out on it. You're hurting somebody who's been there before. We're going to lose in the long run. The way I look at it, 20 years from now it could be really dangerous from uranium mines, acid rain and herbicide spraying. All those three things are harming the land and are very dangerous.

Now I want to talk about the Key Lake mine. Two years before the Key Lake mine opened we had a town meeting with the company. We were promised 65-70 jobs and that the little town of Pinehouse would boom; shopping would take place in Pinehouse and even laundry. But at that time we didn't take minutes of the meeting because we trusted the company. At that time we didn't know much about radiation. Later we got 7 jobs, now there's 3. There wasn't a boom. Then two years later they wanted to open a limestone mine about 12 kilometers from Pinehouse. We let them survey the area but when they wanted to bring in the machines we said no. They had to open a limestone mine in the Hudson Bay area instead.

I've talked to a very experienced miner who's been to all the mines. He says they don't care at all how they handle people there at Key Lake. It's the most careless mine. You can stay as long as you like in the pit. They never tell you to get out. For himself he knows that he has to get out of there every once in a while because of the radiation. But there's people who don't know and nobody tells them to get out. He's worked at more than 20 mines.

The miners that go down in the pit are not being told that they have to get out after a while because of the radiation. They wear badges to measure the radiation but the badges are taken by the company. The miners themselves don't know how to read those

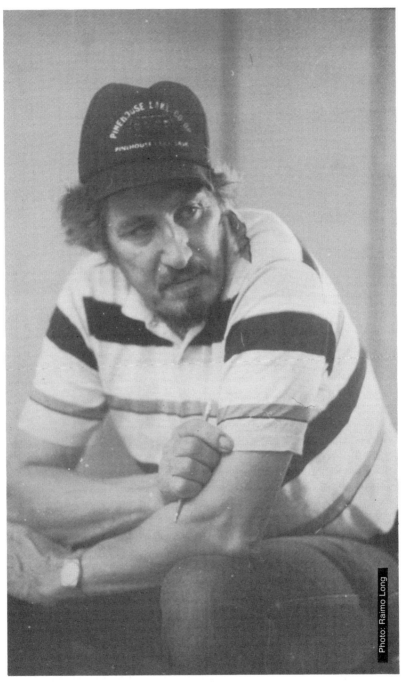

George Smith, Mayor of Pinehouse.

Children on Pinehouse beach.

badges. Just the company knows how much radiation they get. That's no good to the miners.

In Pinehouse we did a planning study to see how much we get from the land - $1 1/2 million per year; $600,000 per year from fish. I'm sure it's the same in other northern communities. The uranium mine can never replace your lake. I think that the government and companies always say they're mining for northern people, but it isn't that way. The Key Lake mine was supposed to give 65% of its jobs to northerners, now it's 19%.

It's the same thing with Prince Albert Pulp. Indian people need the land. It's the land that feeds us. We can't let the land go. Prince Albert Pulp wants to spray herbicides so the spruce can grow. The chemicals are dangerous. Sometimes I think there's going to be no end to this. I'm always going to meetings to try and stop the spraying, and more meetings afterwards.

Look at Beaverlodge Lake where the radiation is too high. There's nobody fishing that lake now, no sport fishermen, no

Pinehouse youth.

domestic fishing. That was a good lake and there's no way that radiation should have went to that lake from the Uranium City mines, but it did.

There's a danger too of the government saying to the fishermen at Wollaston Lake, "If you start saying you'll get radiation in the lake then people won't buy your fish." That's a scary situation.

Things are getting worse because the mines opening up now are more dangerous, like Cigar Lake and Collin's Bay. We have to be really worried. We're much lower educated than the white people outside. It's not easy for us. Us Indian people, if we don't have the land, then we won't survive.

We can't compete with the southern white society. They give us grade 9 education and then we walk the streets and have no jobs. We need the land. Sure we need jobs but I don't think uranium mining or herbicide spraying will give us jobs. They just give us pick and shovel jobs. And what good is uranium anyway? What we need are all kinds of community programs like water and sewers

The road north to the Key Lake mine from the Pinehouse turnoff.

and processing our fish and berries.

The price of alcohol is the same price everywhere. The price of alcohol is the same in Buffalo Narrows as Regina, no different. But a bag of flour is a lot higher in Buffalo Narrows than it is in Regina. If they wanted to discourage drinking then they at least wouldn't subsidize the freight. The best way to deal with the problem is to have more people involved with working and recreation and all that. Get them doing something. In the north here we've got lots of potential to start up many different things, but we don't have a chance. If people were working there would be less drinking.

The government gives out wrong information. They've always been lying. They've got a lot more power and they get a better chance at the newspapers and other media than we do.

Most of the people in the north don't know where the uranium is going. They just figure it's going down to Saskatoon and that's it. There's not enough education on uranium mining.

I think it's good that some people from the south are giving us help, fighting for us. It's not good enough though that only people from the south are fighting. The people from the north have to start speaking for themselves and the more you get the better.

In closing I'd like to say I'm always against uranium mining and herbicide spraying; not for me, I'm an old man, I'm not thinking of myself but the future generations. They need this land.

(Above) Carving at a trapper's cabin near Pinehouse. (Below)
Martin Smith, Pinehouse resident, trapper, and ex-uranium

miner preparing a trap; (Above) with bear skins stretched to dry, and (below) holding wild rice. **Photos: Lillebror**

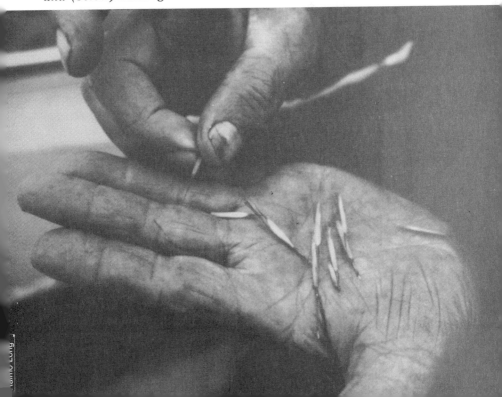

Statement By Brian "Banjo" Ratt, Ile à la Crosse[10]

I'm Cree. We've always been known to be really superb warriors because the Crees stretch right across Canada. But now people are starting to realize that the Dene people are also stretched across, just a little farther north.

A Cree was a diplomat. He could go to any Nation and get along with people. Plus, they had a lot of medicine. They call it voodoo or paganism, but it's coming out really strong right now. The church stifled it. We couldn't talk our own language. We couldn't pray to our creator. We couldn't grow our hair long. We couldn't do all these things because the church stopped it. They built a big boarding school to put all the kids in and try to educate them. The first thing they did with all the boys was shave their heads, then take them to a supper table and expect them to look up at a picture of Jesus Christ who has long hair and a big moustache. It makes a person think and wonder what the hell is going on.

We got sold out. A lot of our own native leaders were forced to get into that development because they had nothing, nothing at all in the north, just welfare and jails. The statistic in the late 1970's that the politicians were using was 75% unemployment. They said we need more money for welfare and that we need to sober up a lot of our people. They actually used that statistic to push uranium mining. They were saying, "Look, look how poor we are, we can't oppose the uranium mines because we're too poor, our people are all poor and living in shacks, we have to have jobs you know." A lot of people were saying, "OK, right on." And they got new houses and things like that. But do you know what percentage of the people are poor today? It's 90 – 95%. What the hell! Who's going to tell me that uranium mining is good. The people are now actually in a worse situation than they were five or six years ago. It's like that in the whole north, not just Ile à la Crosse.

The government is trying to get rid of the traditional people. The whole thing is running away on us. There aren't too many traditional people left. When the majority of the people have become modernized and assimilated, they still shouldn't destroy the traditionalist. It's his government too. But they're judging everyone by

58

what the native leaders are doing. They're letting everyone go down the drain. They're trying to get everybody to assimilate, and the system doesn't have the accommodations for all of us to become lawyers or doctors.

Our water up here isn't perfect but it's beautiful. Down south in Saskatoon you can go to the river bank and see the river. It's green, full of pollution. You don't even want to swim in it or wash your face because you get all itchy and scratched. Up here you can dive into the lake. We've got to really watch our water. Right now we have such a beautiful supply, nice clear water compared to Regina and Saskatoon.

We have to watch what the government and corporations are doing up here. Like there is the Cold Lake tar sands plant just west of Ile à la Crosse. It's not pure oil. It's oil mixed with sand. They have to use a lot of water to mix it up and bring out the oil and throw the water away. The river that comes out of there is called the Beaver River. It flows all the way from British Columbia. We have to watch what is happening because they're trying to take our water away from us. If the water level goes down the animals are affected too. But we could buy six big tankers and haul water down to the south and let them buy our fresh water to drink instead of polluting it.

It's hard for people to comprehend something that's going to happen 20 years from now because they're so worried about trying to make a living today. The governments have people in a position where all they can do is survive. Maybe 20 years from now they'll build a huge cancer clinic, who knows.

The main thing is that people are forced to work at places like the mines because of economic reasons, because of poverty, because they're so poor. And welfare too, they've got you in such a position that by the time you try to do something about it you've got no money to do it, because money makes things go around and there's not much of it.

People are divided right now in Ile à la Crosse. You have all the class struggles that are happening everywhere. Our own native leaders are neo-colonialists now. They are our own oppressors. They don't need radicals like us anymore because they got what they want from the government. Now they don't want us to create problems, they might get their money cut off, so they come after us.

The way I look at it I think that a lot of our own native leaders are really kind of happy that many of their own people are illiterate, because then they can get away with a lot of stuff that people don't know about. When people start getting education the lead-

60

"I don't want to be poor, but I don't want to destroy my own land just to be rich."

ers can't just go ahead and start doing things because someone from the Band is going to question him.

I know I'll never go work at the mine. I don't care how poor I am. I don't want to work at the mine because I believe uranium mining is destroying the land and the people too.

And before we can talk about the bomb we have to talk about where the bomb comes from. It came from the land, from the people. Before they build a bomb they've already destroyed the Indian people. They've destroyed their water and their fishing before they've destroyed people with the bomb. That's the thing we have to look at. We've got to get to where it's happening, right here in the back yard of our country.

There are so many new things thrown at native people, like liquor. That's the first thing the white man brought us. I quit drinking in 1983 because I couldn't handle it. When I drank the first

Photo: Vivianne Johansson

Banjo being interviewed for Finnish radio, August 5, 1985, Ile a la Crosse.

thing I knew I'd be in jail for two years for something idiotic, some-thing so simple that if you were sober you could avoid but when you're drunk your mind is different. Plus, you've got a lot of bitter-ness in your heart and you want to take it out on somebody and the first white person you see is the one you go get – that is for people who hold authority positions like cops and priests.

If my kids don't know how to talk their own language by the time they're 15–16 then I'm the one at fault because I don't speak that language every day. I know that my kids are not going to be able to speak Cree. People have such a defeatist attitude about the whole system.

It's not that the people are for uranium mines but it's that the government doesn't give them the right information. Plus, a lot of times our own leaders are collaborating with the government in order for them to get jobs. They give excuses like we have 90% unemployment in northern Saskatchewan and we have to have the mines for people to work. That's what they told us in 1980. But today you don't see very many people working in the mines. They poured about three million bucks into Ile à la Crosse and everybody just went crazy, just like a mass sellout. But there were a few people saying, "We don't believe in that mine, they're just tricking you."

My father, he's fished and trapped all his life. Within a three month period he'd only make about $3 – 4,000 on furs and fish and he didn't need all that money because he lived off the land. But now you need more money because there are fewer moose now. So my Dad got a job in the mine. After 21 days he came back with $3,400. After 21 days! And here his son is trying to tell him about uranium mining. He just kicks me out the door. But now he's more bitter because he's found out I've always been telling the truth. But it's too late.

I don't want to live poor all my life, but I don't want to destroy my own land just to be rich. There has to be other avenues for people to take. There has to be. Only about eight people work at the mines from Ile à la Crosse out of the whole population of about 1,500. Every week they bring out millions of dollars worth of yellowcake from the mine. Why don't they drop off $2 million right here. That's not happening. They just disregard us.

All around Ile à la Crosse people used to have farms. They had cows and chickens and farmed their own land. Then the socialist government came along and centralized everybody. They said, "We'll give you a brand new house, running water, electricity." A lot of people went for that. They left the land and moved into town. Now they've got nothing. When they want to pay for something they have to have a welfare check to pay for it. That's what the socialists did to the people here in the north.

The houses that they were promised they'll never be able to pay for in their wildest dreams, like $90,000. Some of them are lucky to even get $10,000 a year on welfare and they have to pay for that house out of their welfare. They have to pay for that phone bill, electricity, and water. By the time it comes down to spend that money there's nothing there. They've got the people so poor that by the time they realize what's happening and they try to organize, there's no money there to organize, there's nothing there. Right now the people cannot mobilize without a government grant. The government has made people dependent.

The government has taken all that land and is trying to use it for tourism. They don't want people to go out and fish for their families, they want you to buy fish from the store. You can't go out and put your net in the lake because they're saving all the fish for the tourists who come up from the south with tourist dollars. It's pathe-

The northern lights and the big dipper over the Ile à la Crosse school, September 1985.

tic I think.

We've been talking, talking, talking for years and nothing has changed. It's even getting worse. They're out-manoeuvring the whole anti-nuclear movement. Even the peace movement is just a farce now. Where is the peace movement when you have a Star Wars? It's no use talking any more, we must act. A lot of people only get involved in these movements because they've got a lot of money and a lot of time. Well us poor people we don't have a lot of money and we can't go out and give up our cheques. You want to do that but you've got a wife and two kids to support.

What they do is they divide and conquer because of the almighty dollar. They put us in such a sorry state and then throw us just a little bit of money. The people just rip each other apart for it. It's just like a pack of dogs. You keep them behind a fence and don't feed them for ten days, then you get a nice piece of fresh meat and throw it in the cage. The dogs will go at each other. There'll be ears falling off and the dogs will kill each other for that piece of meat. It's the same thing that's happening now. The government makes you so poor that when they give just a little bit

The Ile à la Crosse grave yard with mass graves from the early 1900's when Cree people died of smallpox. Contaminated blankets were intentionally distributed in the community by the colonialists.

of money everybody starts fighting over it, and there isn't enough money to give everyone a job.

There's a problem with housing too. In 1985 they spent about $300,000 to build just six single family units in a little apartment block. They built them and then said, "OK we want applications from those that want to live here." There was room for six families but 120 families applied. Then the fights started. "I want that house, no I want that house, I got seven kids, to hell with you I got 17 kids and I don't have a job." It's just crazy.

The government comes along and puts a bunch of money in front of the people and says, "We want your land but we'll give you this much money." What the hell are the people going to do with that much money when they've been poor all their lives and lived off the land? They'll go crazy with it. Mismanagement, ripoffs, just outright ripping off your own people. You eat all the meat and there's nothing left but the bones, and people are even fighting over the bones now.

In The City: Alcohol, Welfare, And Jails

The poverty and lack of opportunity in rural areas has pushed many Indian people into the cities. All too often the change does not improve their situation. Canadian jails hold a disproportionate number of Indian people compared to their percentage of the overall population. Though native people make up only a few percent of the national population, they make up over 80% of all prisoners. Alcohol is involved in about 60% of all their offences, and soft drugs in a further 15%.

The following is an interview made in 1985 with a government employee in the prairie region (Alberta, Saskatchewan, and Manitoba) who prefers to remain anonymous. This person has worked for many years in an urban setting with Indian and Métis people in trouble with the law.

Many native people have been brought up on the tradition that everyone is part of you and your family, the great extended family. It is common on the reserve to just walk into any house. If a pot of soup is on the stove and nobody is home, you help yourself, rinse out your dish and leave. Or, if it's pouring rain or snowing, you walk in and go to sleep on the couch. If whoever lives there comes home and sees you, they don't call the police. But do that in the city and you've got three cruiser cars and six policemen crawling all over because you committed unlawful entry. Or, if they don't happen to like you, maybe charge you with break and enter although you had no intent of breaking and entering and committing an offense; you've been brought up on the principle that everybody shares, then you come into the city and nobody shares. There are some people that do but they are in the minority and it's unfortunate.

I think that one of the major reasons why Indian people continue to get into trouble, even without the alcohol problem, is the economic situation. There just aren't enough jobs. They get on the vicious treadmill of, "I haven't got a job, but I've got a dollar, I'll go have a beer." They go down to the bar and run into 10 friends who each have a couple of dollars, and next thing you know they're

66

drunk. Then they're walking through a store or see a car parked and think, "I could take something and make a little more money, get a little more booze." And so they start to break the law. The offense was made to help support the alcohol or drug habit. Most often, if they wouldn't have taken a drink they wouldn't have been in court. It's the alcohol and the economic situation that brings them to the court room in the first place.

The non-native people have to be educated too. I think that one of the reasons that a lot of native people drink is that it's their way of rebelling against the situation they find themselves in. It's like a catch 22, you can't go this way and you can't go that way, there's no one that can help you, there's no-one that has an answer. They're rebelling from being put on a tiny reservation and being told this is where you have to live and given $5 per year treaty money, and free medicine and glasses. If you're not Indian you've got Medicare. So what are they really giving? Not a damn thing. But the politicians convince you they're giving everything.

Indian people don't commit more crimes, but they get charged more often. If a white person, a non-native person went into a large store and took a pack of cigarettes, the security guard might take them into the office and say, "Look don't do that again." But if an Indian takes a pack of cigarettes, it's arm behind the back, call the police and take them away. Not because they don't like Indian people, but because they feel they don't spend much money in the store. That's the way it is, it's a fact of life.

People are prejudiced against minority groups, particularly if their skin is a different color. Indians do get treated differently. If you're Indian with a criminal record and you happen to get picked up, and this is not just unique to the prairie region, this is anywhere in Canada, if you are Indian or part Indian, you're subjected to derogatory remarks by certain police, and subjected to cruel and unusual treatment.

For example, there is the case of a young man who was arrested on impaired driving, he was supposed to be drunk. The police put him in a cell. He was a tall, good looking man with braids past his shoulders, not long enough to wrap around his neck, but long. Next morning he woke up and found that his braids had been cut off. The reason given was that they thought he was going to try and hang himself. In every police station they have special cells for people they think might want to commit suicide. They can handcuff him and chain him to the cell so he can't hang himself. So why cut his hair?

You can talk to anyone you want to and I don't think you'll find too many people that have an answer to the Indian poverty problem, other than to say education, and I don't mean grade 11 and 12 and university. I mean educating them on how to handle their lives, how to handle the alcohol problem and how to handle money, how to be trained and go and look for work.

People are becoming more conscious of the fact that it's important to be able to read and write. When I'm taking information down and ask, "What education have you got?" They say, "I don't have any." They never even learned how to read and write. That shows what one of the problems is. If you can't read and write how can you get a job? If you can't get a job, how can you exist? Eventually you have to put your pride in your back pocket and say, "Hey, I want to learn how to read." That is the first big step.

The government has tremendous programs to create jobs but they're for people who have a good education. They have summer youth positions for students, but the students that get hired are the ones that live at home and don't need to pay board and room.

The government is trying to do something about the problems, but it still just amounts to a bandaid. What we need is major surgery. The government in my opinion doesn't give a damn about poor people. When a single parent mother with one or two children can't work because she has to bring up her babies, does somebody say to her, "We'll provide to look after your children while you go back to school." No, they give her just enough for rent and not enough for food and if she wants anything else it's too bad because she's not going to get it.

If they would have a program for instance that when a mother with a small child can't work because she doesn't have an education, or is sick, or maybe she wants to be like every other woman and bring up her child a little bit first so that the child knows they have a mother, that if they would say, "All right if you would do 10 hours of volunteer work a week, that's 40 hours a month, we'll give you an extra $100 a month." Something like that. She'll feel like she is earning it. She'll feel better about herself and actually be able to afford to go into a restaurant with her child and sit down and say, "I'd like a coke and a hamburger and french fries please." Right now they can't afford that. A lousy little hamburger and french fries! You dream about things that everybody else takes for granted. You shop at the Thrift Stores the Salvation Army thrift store. You buy second hand shoes and clothes so that your children and you are dressed halfway decent because you can't afford to buy things new.

These are the people that the government cuts back on. These are the people that are told, You can't have any more money. That is why when you ask, "Do they have programs for the poor people?", the answer is no. Certain provinces have tried to do something but because all the costs are shared by the federal government, they can't.

I have never met anyone who liked or wanted to be on welfare. And yet people say, "Oh, well that person doesn't try, that person doesn't look for work." How can a girl who has a grade 7 education and a small baby go out and compete for a job when there are university students looking for work. And yet she will go out and do anything. In fact some of them do, you see them on the corner and they charge 50 dollars; not because they like that kind of life but because they want to give their children something or they want to buy a nice dress for themselves just once.

Look at the situation in the prisons. It used to be that if you were in the penitentiary you could take university and electronics courses that the government paid for. If you were doing five years you could take five years of university and get your Master's degree in something. Then, when you got out, there would be no reason for you to break the law anymore. You could get a job. They cut that out to save money. So all they do now is just warehouse people. They just throw them in the penitentiary and say, "Do your time, or go work in the kitchen and we'll pay you a dollar and a half a day."

When they cut programs, if it's a needed program, if it's going to help the little guy, the poor person, that's the first program to get cut. And that's too bad. We have some very good politicians who fight for the little guy but there aren't enough of them. That's it, it's too bad.

Indian people have to be careful to pick Indian leaders who care about Indian people. You can go and have an election but you have to make sure the people aren't just taking the position for the money. Everybody is human. Everyone wants to feather their own nest. You have to make sure that if you're going to be fighting on a political basis that you have people who are going to be doing it because they care and not because they're going to make some money.

Right now we have a Department of Indian Affairs. Indian people that are Treaty Indians are the ones that are handled by Indian Affairs. If you are Métis you are not handled by Indian Affairs. The head of it is not Indian. They've never had an Indian person running Indian Affairs. So how can you have a person who has never been there, who has no feeling in their heart tell you how to run your affairs, how to spend your money, how to develop programs, or whatever else you do?

To some extent the situation has improved. The Bands are starting to get some of their land back but not all of it. It's throwing a few crumbs to save the cake for somebody else. That's about what it comes to. If you come to my house for a meal and I fill you up with soup you think you have been well fed because your stomach is full. Meanwhile, all I gave you was a bowl of soup. I saved the steak for myself. After you leave I'm going to sit down and eat that steak. That is what Indian Affairs is doing in my opinion.

CHAPTER 2
THE MINES

Saskatchewan: The "Saudi Arabia" Of The Uranium Industry

Uranium mining is the "front end," or beginning, of the international nuclear industry. In 1984 Canada became the western world's number one producer and exporter of uranium, surpassing the United States for the first time. The approximately 11,000 tonnes produced in 1984 represented about one-third of the western world's sales. About one-tenth of production is for domestic use.

Uranium from Canada travels to points all over the world. Before leaving the country, most of it is taken to Port Hope, Ontario to be refined, though some goes directly to the U.S. According to official statistics the list of countries that have or are receiving uranium from Canada include: Argentina, Belgium, Finland, France, India, Italy, Japan, Pakistan, South Korea, the Soviet Union, Spain, Sweden, Switzerland, Taiwan, the United Kingdom, United States and West Germany.

All uranium mines and mills have a waste outlet pipe. The one at left is from the Beaverlodge uranium mill near Uranium City. It closed in 1982.

Producing uranium mines in Canada are located in northern Saskatchewan and Ontario. The most lucrative ones are in the homeland of the northern Saskatchewan Dene people. The uranium rich geologic formation known as the "Athabasca Sandstone Basin" covers almost the whole northern third of Saskatchewan.

Most of northern Saskatchewan is still a beautiful area of forests and interconnected lakes and rivers with an abundance of wild berries, other edible plants, wildlife, and fish. But this 300,000 square kilometre area is dotted with radioactively contaminated "sacrifice areas" – a term used to stress that the area around uranium mines remains unsuitable for human habitation for thousands of years.

Three main factors have combined to make Saskatchewan the most important production and growth area for the global uranium industry, as follows:
– low cost of mining due to unusually high grade deposits
– government susidization
– weak public opposition.

High Grade Deposits

Uranium ore normally contains only a few tenths of a percent uranium. In contrast, several large deposits in northern Saskatchewan contain ore grading in the tens of percent. Further, most of the rich deposits are close to the surface, which lowers the cost of getting the ore out of the ground. Many of these high grade deposits are more than 100 times richer than competing mines in the rest of the world. For example, the average grade of the Elliot Lake, Ontario uranium deposits is .1%, while the Cigar Lake deposit in Saskatchewan has an average grade of 15%.[1]

Mining the high grade deposits in Saskatchewan has confronted the uranium industry with uniquely serious environmental and worker protection problems. The first high grade mine to open was Cluff Lake in 1978, followed by Key Lake in 1983 and Collin's Bay in 1985. No other uranium mines in the world have had to deal with such high radiation levels. For this reason, anti-uranium mining advocate Maisie Shiell from Regina, Saskatchewan rightfully calls the new mines "experiments." The high grade Saskatchewan mines have, in addition, all the problems associated with low grade uranium mining.

In 1979 when pockets of 45% ore were being mined at Cluff Lake, the owners bragged that in one day they took out over $9CDN million worth of uranium. It is so profitable to mine uranium in Saskatchewan that the province is known in industry circles as "the Saudi Arabia of the uranium industry."

Bill Harding, a spokesman for the Regina Group For A Non-nuclear Society, was asked in a 1984 interview if it is profitable to mine uranium in Saskatchewan. In his answer he stressed that there are many hidden costs that are not normally considered. He stated,

> To put it in a straightforward and simple way, it is more lucrative to mine uranium in northern Saskatchewan than in most other parts of the world because the ore is richer and the method of extraction is cheaper. But that is only looking at the purely economic considerations that don't take into account the cost of money that the public of Saskatchewan has put up front so that the whole thing can be operated. And it certainly doesn't take into consideration the environmental costs over generations and generations or the costs of social abuse in the north of native people.

Government Involvement

Government subsidization coupled with low engineering costs makes northern Saskatchewan especially attractive economically for multi-national mining companies. For the purchaser of uranium, the price is kept low and government involvement guarantees uninterrupted supply.

Both the provincial and federal governments are directly involved in the uranium industry. The federal government owns Eldorado Nuclear Ltd. (ENL), and the provincial government owns Saskatchewan Mining Development Corporation (SMDC). ENL purchased the Rabbit Lake/Collin's Bay operation from Gulf Minerals in 1982, and is part owner of the Key Lake mine.

SMDC has part ownership in numerous exploration projects, and the Key Lake, Cluff Lake, and Cigar Lake mines. A 1975 revision of the Saskatchewan Mineral Resources Act requires all new exploration and mining projects to offer up to 50% participation to SMDC. The Saskatchewan government is involved in almost every major uranium development in the province.

Government of
Saskatchewan

AMOK LTD.
AMOK LTÉE.

ELDORADO
ELDORADO NUCLEAR LIMITED

Uranerz

Saskatchewan Mining
Development Corporation

Atomic Energy
of Canada Limited

Greetings!

We're the folks controlling the big Saskatchewan uranium boom.

We sell uranium around the world to make nuclear weapons and fuel atomic reactors

We trade in terror — for profits.

The real secret of our success is that we don't discriminate. We peddle uranium and nuclear know-how to anyone. Hell, there's a big demand for the bomb!

We poison the air, water, land and life. That's our business.

The poison that makes us rich makes you poor. What makes us powerful makes you sick.

Can you stop us?

All Power to the Corporations

Unleash the Fury of the Ruling Class

Authorized by: Mutants for a Radioactive Environment,
Association for the Advancement of Atomic Weapons,
and the above-mentioned corporate and state interests.

Produced by: The Committee of 4.5 Billion

Throughout the 1970s the government of Saskatchewan diligently built the roadways and other infrastructure necessary to accommodate the boom in uranium exploration and mining. Resistance to "progress" was discouraged by co-opted native leadership, who emphasized economic benefits to impoverished constituencies. The government sponsored Federation of Saskatchewan Indian Nations proceeded to set up, with corporate/government funding, the Saskatchewan Indian Nations Corporation (SINCO). One of the corporation's main functions, through SINCO Trucking, is to truck uranium from the mines to the distribution hub of Saskatoon. Another special arm of the enterprise, SINCO Security, is responsible for guarding the Cluff Lake and Key Lake uranium mine sites.

There is another important aspect of government involvement in the uranium industry. Since the government is responsible for setting and enforcing health and environmental regulations, and at the same time is part owner of the mines, a "fox guarding the chicken house" situation exists. The result has been lax licensing and environmental regulations by both provincial and federal governments that help encourage new developments and ensure profitable operations.

Weak Public Opposition

The high grade of uranium ore close to the surface plus partnership with Saskatchewan and Canadian government corporations are two of the main factors attracting transnational investment. Just as significant, however, is the relatively weak public opposition in Saskatchewan as compared to many other places. The president of one company operating in Saskatchewan was quoted in the March 1985 issue of *Saskatchewan Business* as saying, "I'd rather face the technical problems of mining Cigar Lake than the political hassle of developing a uranium mine in Australia."[2]

A major reason for the lack of strong public opposition in Saskatchewan is the remoteness of the mines. All are located in the north, far from southern urban centers. To the vast majority of people, the waste problems are "out of sight and out of mind."

International corporations are enjoying the political climate of the Saskatchewan uranium business. The constellation of transnational resource corporations now active in northern Saskatchewan include French, West German, American, and Japanese based interests.

In mid-July 1985 the "most important uranium contract in Saskatchewan's history" as the press called it, was signed between Canada and Japan. SMDC and Eldorado Nuclear Ltd. committed themselves to sending 2.7 million kilos of uranium concentrate to Kyushu Ltd., a Japanese utility company. The delivery is expected to take 13 years, beginning in 1987 and ending in 1999. This quantity is equivalent to about half the annual production of the Key Lake mine.

In a front page article in *The Northerner* (published in La Ronge) on July 24, 1985, Sid Dutchak, Minister in charge of SMDC, stated,

The contract ensures that we will be proceeding with further developments. It definitely signals more discussions and more work on potential developments.

Following are some comments of interest about the six most significant developments: the Uranium City area, Rabbit Lake and Collin's Bay, Cluff Lake, Key Lake, and Cigar Lake.

Closed, Operating and Planned Mines

Uranium City Area

Uranium mining stopped in the Uranium City area in 1982 after 30 years of operation. The population of the community dropped from 4,000 in the 1970s to less than 200 in 1983. Those that remain are mostly the Indian people whose ancestors have always been there. Other than a few small privately owned open pit mines, the whole operation was owned by Eldorado Nuclear Ltd. The Beaverlodge mine, near the Uranium City townsite, is the only underground mine in northern Saskatchewan. A shaft there is almost two kilometres deep.

It is no secret that contamination from the mining had more

The mud flats in the centre of the photo are radioactive wastes dumped from the Beaverlodge uranium mill. The waste outlet pipe, shown at the beginning of the chapter, enters from the right.

to do with mine closure than authorities admit. Wastes from the 25 open pit and underground mines were dumped without treatment onto nearby land and into lakes. In the late 1950s and early 1960s millions of tonnes of solid and liquid wastes were dumped directly into Lake Athabasca. From there contaminants can flow down the Slave River and into the Mackenzie River, which flows into the Arctic Ocean. Contamination continues to reach Lake Athabasca via rain and snow runoff, and ground water flow.

In the 1950's and 1960's when the danger of radioactive contamination was unknown, the fine, sandlike, radioactive wastes (or tailings) were used as construction fill material in Uranium City. The school, most of the streets and buildings, including the hospital, were built on the radioactive sand.

CANDU High, the high school in Uranium City, was named after the Canadian made CANDU nuclear reactor. In 1977, the

federal government agency Health and Welfare Canada found radiation levels from radon gas 60 times higher than their allowed limit in the class rooms. An air venting system was installed to blow the radioactive gas outside, but the vents pointed out into the school ground where the children played.[3]

Rabbit Lake and Collin's Bay

The Rabbit Lake mine operated from 1975 to 1984. The pit is about 550 metres wide and 150 metres deep. The average grade of the Rabbit Lake ore was about .3%, though some pockets of ore graded more than 10 times higher.[4] The Collin's Bay B-zone mine, just six kilometres north, started production in the summer of 1985. Eldorado Nuclear Ltd. owns both operations. The Collin's Bay ore is trucked to the Rabbit Lake mill. One of the mill's notorious problems is that the liquid wastes frequently contain levels of ammonia toxic to fish. The mill produces close to $100CDN million worth of yellowcake every year.[5]

To accommodate the Collin's Bay ore, the mill had to be expanded and altered. During the construction period, a number of workers were fired for exposing poor safety standards and attempting to unionize. After eight months of trying to get their jobs back without success, four workers held a press conference in Saskatoon in January 1984. One of the four, Mike Toft, stated that 52 workers, in May 1983, were abruptly flown off the

The Rabbit Lake pit.

mine site without explanation after working only two weeks. Mr. Toft reported that the workers were:
- exposed to radiation but not provided with personal radiation monitoring badges as required by Health and Welfare Canada
- forced to stand ankle deep in water contaminated with radiation,
- exposed to radioactive dust from an ore stockpile, rock crusher, and mill concentrator
- working perilously close to where explosives were being loaded. At one point a blasting error showered the construction area with flying rock.

To try and improve the situation, the employees applied for trade union representation. As soon as this was discovered by Eldorado Nuclear Ltd. the workers were fired, and the company they worked for, Enerpet Construction Ltd. of Calgary, lost the contract.[6]

The Collin's Bay open pit mine is especially dangerous because the uranium is actually under the bottom of Wollaston Lake. In order to get at the uranium, part of the bay was diked off and drained in 1984. Mining below the bottom of the lake began in the spring of 1985. The pit is separated from the rest of

81

The Collin's Bay pit, August 1985.

the lake by a dike of thin steel that extends only about one metre above water level, and may not be high enough to withstand strong waves which are a common occurrence on the lake. After the projected six years of mining the dike will be destroyed, allowing the further spread of contamination and adding to the threat of destroying the commercial fishing potential of Wollaston Lake.

The B-zone is only one of many high grade deposits in the Wollaston area. In the summer of 1985 Eldorado Nuclear Ltd. distributed a glossy color booklet stating:

> When the Collin's Bay deposit is eventually depleted ore *will* be mined from several deposits within a 12 kilometre distance. (Emphasis added.)

Four of these deposits (A-zone, D-zone, Eagle South, and Eagle Point) are all larger than the B-zone and deeper under

82

Containers holding radium wastes at the Cluff Lake mine.

Wollaston Lake.

A number of other deposits are located west of Wollaston Lake in its drainage basin. Some of them are Dawn Lake, Midwest Lake, McLean Lake, Raven-Horseshoe, West Bear, McCarthur, and Cigar Lake. No environmental studies have been done to determine the effect of all these mines put together.

Cluff Lake

At the Cluff Lake mine, which began operation in 1980, extremely dangerous radioactive radium wastes have been placed in ten-tonne concrete containers. After the first ore body had been mined out in 1983, there were 2,916 filled containers sitting at the mine site. Though they were expected to last at least 100 years, some cracked and began to leak already in 1983. By the fall of 1986 more than 200 containers had either cracked or tipped over (they were stacked two-deep), spilling more than 2.5 tonnes of radioactive sludge.[7] In 1982 when there was a spill of about 2 tonnes of the radium wastes, radiation levels were reported to be 600,000 times what is allowed by the regulations.

The Cluff Lake Mining Corporation has not determined what will be done with the radium wastes in the long-term. There are, however, plans to begin extracting gold and uranium from the radium wastes in early 1987. There was no public participation in the decision. At an August 13, 1986 press conference the company announced a new facility at the mine site will be built at an expected cost of $2.6CDN million. It was further stated that the reprocessing is estimated to yield up to 283,000 grams of gold worth about $4.5CDN million, and 56,700 kilograms of yellowcake, bringing in an additional $3.3CDN million.[8] [9] See the "Uranium=Bombs" section for information on ownership of the mine.

Key Lake

The open pit at Key Lake is the largest uranium mine in the world. The owners of Key Lake Mining Corporation (KLMC) are Eldorado Nuclear Ltd. (1/6), Uranerz of West Germany (1/3) and SMDC (1/2). In early 1985 recoverable reserves were estimated at 84 million kilograms at an average grade of 2.5%. In 1984 the Key Lake mill produced 4.7 million kilograms of yellowcake, which was 12% of western world production.[10] The mill is capable of producing about 5.5 million kilos per year.[11] The mill wastes are known to contain arsenic levels that present a serious hazard.[12]

The Key Lake mine has been widely publicized, internationally, as using state of the art technology and being the most modern and safest in the world.[13] Why have public relations efforts been so intensive? Part of the answer is that both the Canadian and Saskatchewan governments had to justify their huge investment. The mine began production in October, 1983 when there was a glut of uranium on the international market, which has continued to the time of this writing. Another part of the answer is the intense controversy generated by concerned people.

The public inquiry into the mine was so overtly biased that it prompted a boycott by all public interest groups. At about the same time there was a court case by Maisie Shiell, an independent citizen, that concluded with KLMC being found guilty of illegally draining a number of lakes. The court gave the company a small fine and permission to continue work.

Further, in the summer of 1981 the Saskatoon based Group For Survival, a group of native people, occupied the site of a planned limestone mine about 300 km south of Key Lake. The limestone was intended for use in the mill as a neutralizing agent. The community of Pinehouse is only 12 km north of the limestone deposit. Pinehouse has a population of 600 Métis people, and is one of the villages closest to the Key Lake mine. The occupation led to the community of Pinehouse opposing the project. Finally, the site was abandoned and limestone had to be transported at considerably greater cost from near Hudsons' Bay.

The only hope the mine owners had of saving face was to mount a massive public relations campaign. KLMC hired a full time public relations staff that coordinated production of glossy, color booklets, a slide presentation, film, and numerous other efforts. In Saskatchewan the company ran full page newspaper advertisements and short TV spots just before the evening news, to catch the maximum possible number of viewers. One TV spot focused on an attractive, young, female geologist working in a sunny, pleasant, forested area.

Much to the embarrassment of KLMC, within the first three months of operation at least 12 major spills of radioactive wastes occurred. The largest was in January, 1984 when over 100 million litres of radioactive liquid with radiation levels at least 20 times the regulation level spilled over the retaining walls of a holding pond.[14] Plans to clean up the mess did not begin until after a one day blockade of the mine road by local native people and supporters. The controversy continues, and so does the PR.

Cigar Lake

The most significant uranium deposit ever discovered is at Cigar Lake adjacent to the southwest shore of Waterbury Lake. It is located 115 kilometres northeast of the Key Lake mine and 55 kilometres west southwest of the Rabbit Lake mine. The main Cigar Lake ore body is the world's largest high grade deposit. It contains over 100 million kilos of uranium at an average grade of 15%, with pockets as high as 60%. This is twice as big and 6 times as rich as the Key Lake "monster deposit." In addition, potential reserves at Cigar Lake are estimated to contain a further 50 million kilos at a grade of 4.7%. Though the deposit was discovered in 1981, its existence was not made public until 1984.

The partners in Cigar Lake Mining Corporation (CLMC) are Cogema Canada Ltd., Montreal (32.625%), SMDC (50.75%), Idemitsu Uranium Exploration Canada Ltd., Calgary (12.875%) and Corona Grande Exploration Corporation (3.75%). Cogema Canada is a subsidiary of the Atomic Energy Commission of France which is directly responsible for France's regular nuclear weapons testing in the South Pacific.

Jim Hasty, a pilot living in Wollaston, does most of the flying for the Cigar Lake project. In his single engine Otter he flies the high grade core samples together with people and groceries from the project site to Barge Landing (Highway 109, across the lake from Wollaston). According to him, CLMC uses the same truck to haul high grade radioactive core samples south that brings groceries north.

Since the Cigar Lake deposit is 410 to 440 metres below the surface an underground mine is being considered. However, no method presently exists to overcome the radiation problem presented by underground mining of such high grade ore. According to CLMC, robots may be needed to remove the ore because radiation levels will be so high in the underground mine shafts. If an open pit is constructed, it would have to be about 2 kilometres long. The deposit has been estimated to extend for 1,850 metres with a maximum width of 100 metres.[15]

Despite the technical and political controversy of the Cigar Lake mine the government has given the green light to the project and production is expected to begin in the early 1990's. In May 1985 CLMC stated they had begun work on a two year program expected to employ 30 people and use a budget of $50CDN million.[16] There has been no public involvement in the decision to mine the Cigar Lake deposit.

Greg R. Land, 1982

Uranium=Bombs

Before we can talk about the bomb we have to talk about where the bomb comes from. It came from the land, from the people. Before they build a bomb they've already destroyed the Indian people. They've destroyed their water and their fishing before they've destroyed people with the bomb. That's the thing we have to look at.

– Brian Ratt, Ile à la Crosse, August 5, 1985.

Ever since the beginning of the global nuclear industry Canadian uranium has played an important role. Uranium from the Northwest Territories helped produce the nuclear bombs exploded over Hiroshima and Nagasaki. Today, uranium mined in Saskatchewan directly feeds the nuclear fuel chains and weapons programs of several western nations, foremost the USA, England and France.

The question of the connection between Canadian uranium and nuclear weapons is not "is the uranium ending up in nuclear weapons?", but "how directly, and how much?" Until the early 1970's almost 90 percent of Saskatchewan uranium went to the United States. At that time there was no doubt that the U.S. was using the uranium for its nuclear weapons, as they had not yet begun a vigorous commercial reactor program.

By the late 1970's the U.S. had found enough uranium within its borders to easily meet the needs of its nuclear power plants. However, Canada continues to export uranium to the U.S. The imported uranium is being used to either replace U.S. uranium intended for reactor fuel so that the U.S. supplies can be used for weapons, or being directly used in weapons production itself. In both cases support is being given to weapons production and testing.

Another clear weapons connection with Saskatchewan uranium is via exports to France. The uranium received by France from the Cluff Lake mine directly contributes to French weapons production and testing capacity. Cluff Lake is owned 20 percent by Saskatchewan Mining Development Corporation (SMDC) and 80 percent by AMOK, a consortium of French government companies. AMOK is owned 30 percent by the French Commissariat de l'Energie Atomique (CEA) which manufactures and

88

tests nuclear weapons.

CEA operates the "Centre of Experiments" in the South Pacific, where regular nuclear bomb tests have taken place since the mid-1960's. The Australian government stopped exports of uranium to France in 1983 because of its bomb testing. After a change of government in Australia, exports of Australian uranium to France began again in August 1986. Canada has left the door wide open for France.

The use of uranium for weapons cannot be separated from "peaceful" uses in commercial nuclear reactors. One of the main reasons is the mutual dependence of the civil and military nuclear industries on each other. Some of the same processing facilities are used in the production of nuclear fuel for civil nuclear reactors as in the production of the two key explosive components of nuclear bombs, plutonium and uranium. Figure 1 shows how the nuclear fuel chain and the nuclear weapons chain overlap in the United States. For other nations, the locations of each step in the process are different but the links are the same.

Plutonium is produced by uranium fission, which occurs in all nuclear reactors. Modern nuclear bombs have a plutonium "pit" and a plutonium or uranium-235 "spark plug," both of which are surrounded in a layer of uranium-238 metal to increase the bomb's destructive capability (see Figure 2).[17]

Canadian uranium even contributes to the USSR's production capability of nuclear weapons and uranium ammunition. Some of the Saskatchewan uranium purchased by German and Finnish companies is sent to the USSR enrichment plant in Riga, Latvia. The enriched uranium is sent on to Germany and Finland, but the depleted uranium (also called uranium-238 metal) stays behind.

Despite these clear connections of Saskatchewan uranium to nuclear weapons, the mining companies persist in stating they do not exist. For example, a May 1985 letter from Eldorado Nuclear Ltd. to Wollaston Lake residents reads:

Contrary to allegations often made by the industry's critics, uranium mining in Saskatchewan is not related to the nuclear arms race. For the past 20 years, it has been illegal to sell Canadian uranium for use in nuclear weapons. Since 1965, the only market for Canadian uranium has been domestic and foreign utilities which use the uranium to produce electricity.[18]

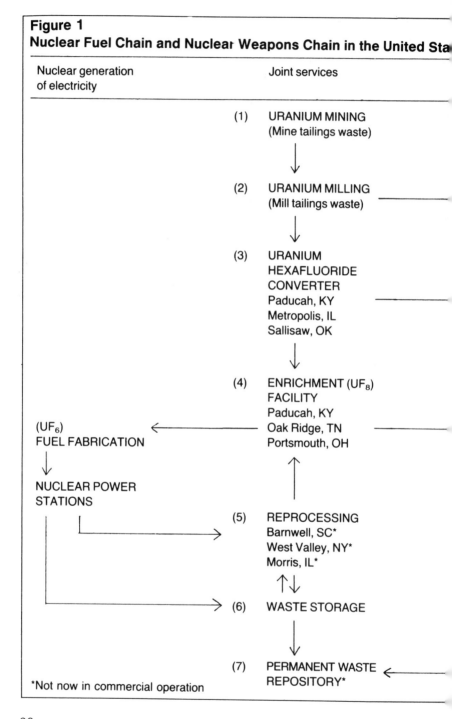

Figure 1
Nuclear Fuel Chain and Nuclear Weapons Chain in the United Sta

Nuclear generation
of electricity

Joint services

(1) URANIUM MINING
(Mine tailings waste)

(2) URANIUM MILLING
(Mill tailings waste)

(3) URANIUM
HEXAFLUORIDE
CONVERTER
Paducah, KY
Metropolis, IL
Sallisaw, OK

(4) ENRICHMENT (UF_8)
FACILITY
Paducah, KY
Oak Ridge, TN
Portsmouth, OH

(UF_6)
FUEL FABRICATION

NUCLEAR POWER
STATIONS

(5) REPROCESSING
Barnwell, SC*
West Valley, NY*
Morris, IL*

(6) WASTE STORAGE

(7) PERMANENT WASTE
REPOSITORY*

*Not now in commercial operation

90

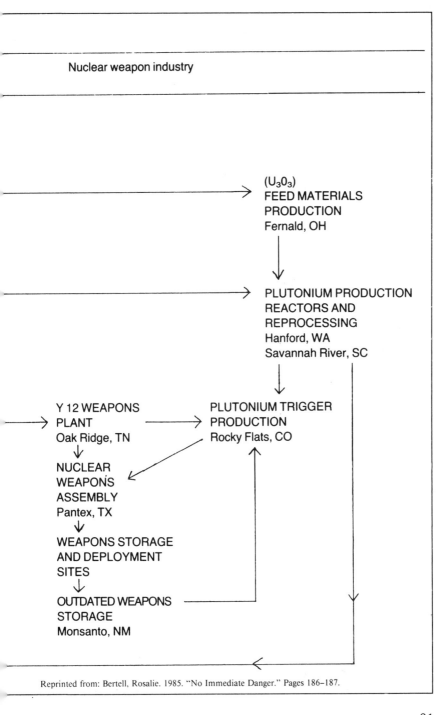

Nuclear weapon industry

(U_3O_3)
FEED MATERIALS
PRODUCTION
Fernald, OH

PLUTONIUM PRODUCTION
REACTORS AND
REPROCESSING
Hanford, WA
Savannah River, SC

Y 12 WEAPONS
PLANT
Oak Ridge, TN

PLUTONIUM TRIGGER
PRODUCTION
Rocky Flats, CO

NUCLEAR
WEAPONS
ASSEMBLY
Pantex, TX

WEAPONS STORAGE
AND DEPLOYMENT
SITES

OUTDATED WEAPONS
STORAGE
Monsanto, NM

Reprinted from: Bertell, Rosalie. 1985. "No Immediate Danger." Pages 186–187.

91

Figure 2

A Nuclear Bomb

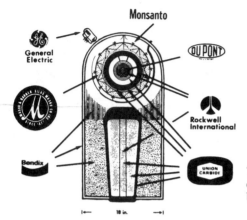

Monsanto

General Electric

Rockwell International

Bendix

UNION CARBIDE

The corporations that make the essential components of a nuclear bomb.

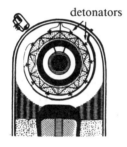

detonators

Monsanto makes the electrically fired detonators surrounding the primary. . .

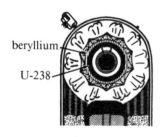

beryllium

U-238

which set off the chemical high-explosive charges, made by Mason and Hanger-Silas Mason, that surround a hollow spherical tamper made of beryllium and uranium-238. The tamper, manufactured by Union Carbide, is liquified by the implosive shock wave and driven inward toward the softball-sized fissionable core of the primary.

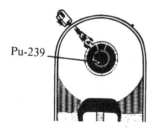

Pu-239

The core is compressed to supercriticality by the tamper, and a beam of high energy neutrons is fired from outside the casing by a high-voltage vacuum tube made by General Electric. The neutrons start a fission chain reaction in the plutonium-239 "pit" made by Rockwell.

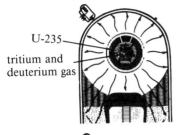

U-235

tritium and
deuterium gas

The chain reaction spreads to a layer of uranium-235 surrounding the pit, and the heat and pressure of fission ignite a hydrogen fusion reaction in the "booster" charge of tritium and deuterium gas, supplied by Du Pont. Fusion adds neutrons to the fission reaction, speeding it up and raising its temperature.

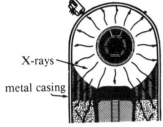

X-rays

metal casing

The energy of fission reaction races away from the primary in the form of x-rays which are momentarily tapped by the bomb's metal casing. . .

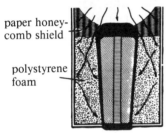

paper honeycomb shield

polystyrene foam

focused through a paper honeycomb shield, and absorbed by a special polystyrene foam "channel filler" made by Bendix, which serves as a thermal explosive encasing the secondary.

The exploding styrofoam compresses the secondary, which is filled with lithium-6 deutride. A "spark plug" of uranium-235 or plutonium-239, running down its centre, is compressed to supercriticality, and a second fission chain reaction thus begins to supply neutrons which convert lithium-6 into tritium.

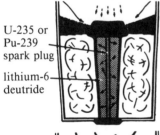

U-235 or
Pu-239
spark plug

lithium-6
deutride

The nuclear explosion of the spark plug generates the temperatures and pressures needed to fuse the newly created tritium with deuterium, showering the casing of the secondary with high-energy neutrons created by fusion. The neutrons cause uranium-238 in the casing of the secondary (called the "pusher") to undergo fission. The lithium and uranium parts are made by Union Carbide.

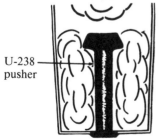

U-238
pusher

The fission of uranium-238 supplies most of the energy that allows an explosive device the size of a kitchen garbage can to destroy an entire city, along with its outlying suburbs.

Reprinted from: Morland, Howard. 1981. "The Secret That Exploded." Random House, N.Y., N.Y., U.S.A 289 pages. See pages 277-279. ISBN: 0-394-51297-9.

Uranium Bullets[19]

Nuclear bombs are not the only military weapons containing uranium. Since the early 1970's depleted uranium (DU) has been used for armor piercing incendiary ammunition. DU is also used in bullets meant to hit people directly. This information has been a closely guarded military secret. Both Canadian and Australian government Royal Commissions into the effects of uranium mining, which tried to be comprehensive in their examination of the uses of uranium, did not make public the use of depleted uranium in ammunition.

The primary purpose of the uranium processing stages, milling, refining, and enriching, is to extract the uranium-235 for use in nuclear weapons and reactors. Uranium enters an enrichement facility containing less than 1% uranium-235; almost all the remaining part is uranium-238. The final product may contain anywhere from 3 to 90% uranium-235. This product is called "enriched uranium," as its quantity of uranium-235 is increased. The material remaining is called "depleted uranium" because most of the uranium-235 is taken out. DU is almost pure uranium-238. It has few uses, exists in large quantities, and cannot cause an atomic explosion by itself.

Military secrecy and industry competitiveness make it difficult to make an accurate estimate of how much DU has been used in ammunition production. However, the US military alone consumed a minimum of close to 10,000 tonnes by 1987. Though this represents a small portion of all the uranium mined in the US (about 300,000 tonnes up to 1987), it is not insignificant. About 10 million tonnes of solid radioactive waste had to be dumped somewhere to produce 10,000 tonnes of DU. Also, this quantity of DU represents an amount of yellowcake equal to about half the total production over the entire 30 year life of the Beaverlodge uranium mill at Uranium City.

Armor Piercing And Incendiary

Depleted uranium is used in bullets because it:
- has a high density; it is the heaviest non-manmade substance on Earth
- it is relatively soft compared to other metals
- it is pyrophoric (starts on fire spontaneously) when finely

94

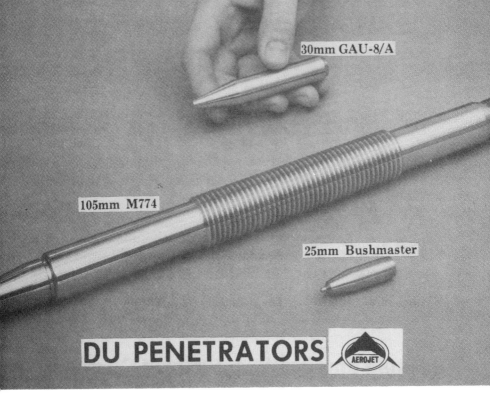

30mm GAU-8/A

105mm M774

25mm Bushmaster

DU PENETRATORS

AEROJET

Reprinted from: Aerojet Ordnance Co. "Depleted uranium Penetrators." Page 20.

One 105 mm DU core in the M774 shell weighs about 3.6 kilos (8 pounds). Several hundred thousand were produced by 1984. The 30 mm projectile weighs 425 gm (15 ounces).

divided.

Ammunition is specially made to take advantage of one or more of these three attributes. DU is also attractive because it is cheaper and more accessible than alternative substances, such as tungsten.[20]

Because of its high density, bullets made out of DU are more efficient than any other material at passing through steel. DU is not only the best armor penetrator, but is required to penetrate modern armor plating.

On impact with steel, pieces of superheated DU flake off and start on fire. Thus, DU bullets can pierce steel plating, come through the other side partly on fire, and ignite any ammunition or fuel it comes into contact with. This has the advantage of causing greater destruction of a hit. What is more, in the process of burning, toxic smoke is formed which can cause delayed effects if inhaled.

95

Standard NATO Armament

DU ammunition is part of standard NATO armaments. In mid-1985, Major General Peter G. Burbules of the US Army wrote to the Nuclear Regulatory Commission in Washington, DC supporting an application by Sequoyah Fuels, Inc. to build a UF4 conversion plant. Uranium metal is produced from UF4. The General wrote that there was an "urgent need for penetrator munitions" and that depleted uranium metal "has certain unique properties which make it a vital component for our defense programs."[21]

However, in some NATO countries there has been a controversy over placement of the uranium ammunition on their territory. In September 1984, the Danish Ministry of Defense initially refused, then allowed placement of the uranium bullets at the NATO base in Jylland, Denmark.[22] The Canadian military stocks DU bullets, and has tried to open a production factory in Canada.

Single and Multiple-barreled Cannons

The United States military uses DU in armor piercing incendiary ammunition fired from single and multiple-barreled cannons on land, water, and in the air. The cannons are made to fire in bursts measured in seconds or fractions of a second. Multiple-barreled cannons are able to fire up to 6,000 rounds per minute (100 rounds per second).

Five different sizes of DU ammunition are made: 20, 25, 30, 105, and 120 mm. The 20 mm is fired from the M61A1 Vulcan six-barreled cannon; the 25 mm from the five-barreled GAU12U cannon and the single-barreled M242 chain gun, the 30 mm from the four-barreled GAU13A and the seven-barreled GAU8A cannons; and the 105 and 120 mm shells are fired from a number of the US Army's tanks, including the main M1 battle tank. The multiple-barreled cannons named are produced by General Electric (GE) in Vermont, and the M242 is made by Hughes Helicopters in California. The 20 mm ammunition is alloyed with 2% molybdenum, and the other rounds with .75% titanium.

Two US corporations manufacturing the different types of DU ammunition are Honeywell and Aerojet. In the mid-1980's, for just the GAU8 ammunition, they shared a $1.3US billion contract to produce 100 million rounds.

Aerojet Ordnance Company claims they are, "a leader in the

96

As part of the A10 program of the US Air Force, Aerojet produced the first 30 mm DU bullets in 1975 and began mass production in 1977. The A10A Thunderbolt II close-support aircraft (above) has a GAU8A cannon mounted in the nose, which can fire up to 70 rounds per second.[27]

military application of heavy metal." Their research and development includes a cluster bomb where small DU bomblets fall out of a dispenser and rain down on a target area. Aerojet has a special branch called "Heavy Metals Division" that has its headquarters near Jonesboro in northeastern Tennessee.[23]

Radar Controlled Cannon

The MK 15 Phalanx Close-in Weapon System uses a radar controlled six-barreled 20 mm Vulcan cannon. It can shoot up a cloud of DU bullets at a rate of 3,000 per minute that will destroy any incoming objects. The system is specially designed by the US Navy to act as a defense against sea skimming, cruise-type missiles. The weapon increases a ship's invulnerability, thus making it more dangerous, forcing "the other side" to escalate its weapons development in defense.

DU was chosen as ammunition material for the Phalanx system after a testing process that lasted about a year. In addition, handling requirements of the US Nuclear Regulatory Commission had to be met. The first Phalanx system was completed in 1979. A ceremony marking production of the 300th was held by the US Navy in 1984.

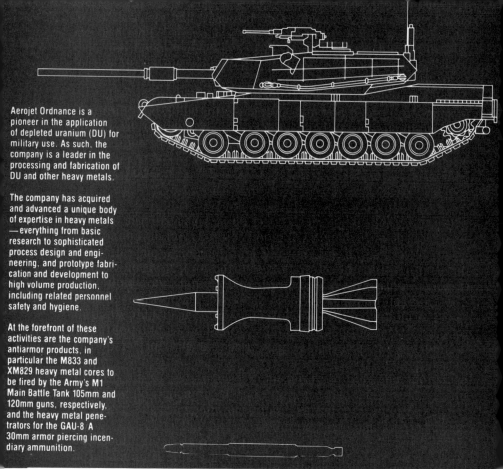

Aerojet Ordnance is a pioneer in the application of depleted uranium (DU) for military use. As such, the company is a leader in the processing and fabrication of DU and other heavy metals.

The company has acquired and advanced a unique body of expertise in heavy metals —everything from basic research to sophisticated process design and engineering, and prototype fabrication and development to high volume production, including related personnel safety and hygiene.

At the forefront of these activities are the company's antiarmor products, in particular the M833 and XM829 heavy metal cores to be fired by the Army's M1 Main Battle Tank 105mm and 120mm guns, respectively, and the heavy metal penetrators for the GAU-8/A 30mm armor piercing incendiary ammunition.

Reprinted from: Aerojet Ordnance Co. Undated promotion booklet.

Anti-personnel Uranium Bullets

DU bullets are soft enough that when they hit human flesh they spread out; thus entering at a tiny point but leaving a big hole on the other side. The military refers to this effect as an "explosive type wound." In order to enhance the damage done, bullets have been made out of 2 cm long needle-like flechettes. The flechette curls over into a hook shape on impact, thus maximizing the explosive effect, and may have a split tip to further increase wounding power. They may be made out of steel, DU, or other metals.[24][25]

Flechettes are used in rockets and rifle and shotgun shells. Flechette cartridges have been made for the American M14 7.62 mm rifle and the M16 5.56 mm rifle. This type of ammunition has also been made for pistols by a French manufacturer.[26]

What About The Wastes?

It is not possible to mine uranium without producing wastes that have a catastrophic effect on the immediate surrounding environment, especially the downstream area. Uranium mine and mill wastes decrease water quality to such a degree that aquatic communities may be completely eradicated in the immediate downstream vicinity. The main forms of waste include: the overburden material; ore not rich enough to mill; mill process chemicals and water; and contaminated clothes, tools and buildings. The following discussion, however, focuses on wastes that come out the discharge pipe from a uranium mill – the water, crushed rock and process chemicals.

This part is divided into six sections. The first is an overview of the nature of the waste problem, presented to establish the basic characteristics of mill wastes. Described are a few aspects of radioactivity, the large volume of wastes produced in a short time and why wastes remain toxic for thousands of years – forever in human terms. Included as well are the often overlooked problems of heavy metals and process chemicals.

The second section describes some effects of radioactivity on plants and animals, particularly the topics of biological accumulation and radioactivity transfer to people. The third section outlines some reasons why there is no solution in sight to eliminating contamination by wastes. The problems of spills and revegetation of waste piles are discussed. In the fourth section some possible remedial actions to stop the pollution are listed. Some precautions are mentioned that can be taken to limit exposure to people from radioactivity in contaminated food.

The fifth section uses the Rabbit Lake/Collin's Bay operation as an example to illustrate how uranium mines do not meet conditions set by their Canadian government operating license. Quantities of lead, arsenic, uranium, and radium downstream from the mine are used to point this out. The last section reviews the position of the authorities and Eldorado Nuclear Ltd. regarding pollution from the Rabbit Lake/Collin's Bay mines.

The Nature of the Problem

Uranium is the heaviest non-man-made substance. Heavier are some radioactive byproducts formed by the atomic reaction inside a nuclear reactor and by the explosion of nuclear bombs.

Uranium atoms were formed billions of years ago by supernovas, or exploding stars. The uranium then became part of the dust forming our solar system and the Earth. Most uranium deposits in Saskatchewan are ancient seabeds. Since uranium is heavier than water, it slowly accumulated over the millennia in the bottom sediment of water bodies.

Uranium deposits exist today because of their isolation from oxygen and water for millions of years. Uranium mines break this natural containment, allowing water and air to carry contamination throughout the environment. In its natural state, uranium, other radioactive materials, and heavy metals are in the form of solid rock and therefore only tiny amounts, if any, can escape to the surrounding environment.

Once the ore is crushed down to a sand in a mill, its volume is greatly increased. Then, the solid wastes are dumped onto the surface and allowed to mix with air and water, entering into a complexity of biological pathways and spreading contamination far from the mine site. For example, radon gas that escapes from tailings piles is essentially isolated from the biosphere prior to mining. Radon gas can be carried at least a thousand miles by the wind and affect large numbers of people.[28]

Despite these facts industry supporters often state that it is safer to mine and take away the uranium than leave it in the ground. At a community meeting in Wollaston Lake on April 17, 1985 Bernie Zgola of the AECB Uranium Mine Waste Management Division said,

An important thing to note is that once the uranium is mined out of the pit the uranium will be gone from the pit, the Collin's Bay pit, it will no longer be there as it is in it's natural state. So perhaps you could look at it from that point of view that the situation will be improved.

Further, this attitude of Mr. Zgola does not consider the destruction where the uranium is taken to. Dangerous wastes are created at every link in the nuclear fuel chain.

The Rabbit Lake mill and accommodation facilities.

Photo: Lillebror

The Milling Process

Uranium ore contains only a few tenths of a percent uranium, except for high grade deposits. All the rest of the rock is considered waste. To extract the small quantity of sought-after uranium, the ore is taken out of the ground and processed in a mill. Uranium mills produce two things – a marketable product, and wastes. The marketable product is a fine, sand-like, yellow material called ammonium diuranate or uranium oxide (U_3O_8), though it is generally referred to as yellowcake. Yellowcake consists of up to 90% natural uranium.

At a uranium mill, the rock is crushed, ground down to a fine sand and mixed with large amounts of water and chemicals. The chemicals are either acids or bases, depending on the pH of the ore. With the exception of most of the ore treated by Eldorado's Beaverlodge mill, all Canadian ores have been and are treated by an acid process. Both of the processes are able to remove about 90% of the uranium but only a few percent of the other radionuclides. About 85% of the total radioactivity in the rock goes out the end of the waste outlet pipe.

101

Radioactivity – Dangerous Forever

Uranium is constantly changing into other radioactive materials that are always present wherever uranium is found. The change from one radioactive material to another is called radioactive decay. Figure 3 shows the decay series for uranium-238. In its natural state, uranium is made up of over 99% uranium-238, less than 1% uranium-235 and less than .01% uranium-234. Uranium-235 has its own decay series.

Uranium-238 changes 14 times before it becomes non-radioactive lead. In addition, there are 22 other naturally occurring radioactive materials from separate decay series.[29] Thus, there are always 36 different radioactive materials in the ground, not just "uranium." Most of them are ignored by the uranium industry. However, from a health perspective they are all important.

The time it takes for one radioactive material to turn into another is measured by the concept of "half-life," which is the time it takes for half an amount of a radioactive material to decay into the next material (called a decay product). In Figure 3, the half-lives are written to the right of   each radioisotope. For example, for 1,000 grams of radium-226 (the fifth decay product) it takes 1,600 years for half of it, 500 grams, to decay into radon-222. There are thus 500 grams of radium-226 left. It then takes a further 1,600 years for half of that, 250 grams, to decay, and so on. An individual atom, however, may decay instantly or take thousands of years. Radioactivity is released every time a radioactive material changes to the next material in its decay series.

Some very long-lived radioisotopes are dumped into the environment from a uranium mill. For example, thorium-230 has a half-life of 80,000 years. It is long half-lives like these that make uranium mill wastes stay radioactive so long as to be considered forever in human terms.

Radioactive particles pose the greatest threat to human health when they are inhaled or ingested. But they are so small that they can enter into the skin via the many sweat pores and hair follicles all over the body. The radioactivity can be of three types, alpha, beta, and gamma.[30] Alpha radiation is the most harmful to living cells but travels only about a centimetre in air. Beta radiation travels about half a metre in air and cannot go through thin steel or wood about 5 centimetres thick. The difference between alpha and beta particles is like a cannon ball compared to a bullet. Alpha particles, like cannon balls, have less penetrating power but more impact.[31] Gamma radiation is the least

Figure 3

Main Characteristics Of The Uranium -238 Decay Series*

Isotope	Half-life
U^{238}	4.5 billion years
$\downarrow \alpha$	
Th^{234}	24.1 days
$\downarrow \beta$	
Pa^{234}	1.18 minutes
$\downarrow \beta$	
U^{234}	248,000 years
$\downarrow \alpha$	
Th^{230}	80,000 years
$\downarrow \alpha$	
Ra^{226}	1,600 years
$\downarrow \alpha$	
Rn^{222}	3.82 days .... Radon Gas
$\downarrow \alpha$	
Po^{218}	3.05 minutes
$\downarrow \alpha$	
Pb^{214}	26.8 minutes
$\downarrow \beta\gamma$	
Bi^{214}	19.8 minutes
$\downarrow \beta\gamma$	
Po^{214}	0.00016 of a second
$\downarrow \alpha$	
Pb^{210}	21.3 years
$\downarrow \beta$	
Bi^{210}	5.01 days
$\downarrow \beta$	
Po^{210}	138.4 days
$\downarrow \alpha$	
Pb^{206}	Stable

Short Lived
Radon
Daughters

Abbreviations:

α – alpha Bi – Bismuth Ra – Radium
β – beta Pb – Lead Rn – Radon
γ – gamma Po – Polonium Th – Thorium
  Pa – Protactinium U – Uranium

*Adapted from: McGraw-Hill. 1977. "Encyclopedia of Science and Technology."
Vol. 11. See page 292. McGraw-Hill Book Company, N.Y., N.Y.

harmful but can travel great distances. The majority of gamma radiation is stopped by a few centimetres of lead or about 30 centimetres of concrete.

Radium And Radon Gas

Radium is one isotope in uranium mill wastes that is especially dangerous. This is because it is known to be harmful to life forms at low concentrations and it decays into the even more dangerous radon gas. Chemist Marie Curie discovered radium in the early 1900's. She and her daughter both died from their exposure to radiation.

The radium problem is particularly serious where wastes have been dumped on stream bottoms as radium accumulates in the sediment. About 99% of the total radium-226 in uranium ore is discharged in the waste from a uranium mill. The Rabbit Lake mill discharges about 1 1/4 grams of radium-226 per day of operation.[32] This amount of radium gives off almost 50 billion radioactive disintegrations per second.

Radon gas has been studied in detail. There are three main reasons why radon is so dangerous. First, because it is a gas and can thus be breathed into the body. Radon is the only gas that occurs in the uranium decay series. Otherwise, the materials change from one solid to another.

Unnaturally large amounts of radon gas are continually coming out of the ground at uranium mine and mill waste areas. A research team at Los Alamos Scientific Laboratory, University of California, studied this problem and came to the following conclusion:

Our research indicates that 4 metres of clay are required to reduce radon exhalation by 99% and the remaining 1% is still about four times the typical soil radon exhalation rate. Perhaps the solution to the radon problem is to zone the land in uranium mining and milling districts so as to forbid human habitation.[33]

The second reason radon is so dangerous is because it releases the most harmful type of radioactivity – alpha radiation. The third reason is that radon has a short half-life, and is followed by the extremely hazardous "radon daughters," or the decay products that immediately follow from the "parent" radon. The first four radon daughters have in total a half-life of less than one hour. Two of them give off alpha radiation and the other two beta and gamma radiation. Once inside the lungs radon decays rapidly, exposing sensitive lung tissue to deadly radiation. That is why lung cancer is so common among uranium miners. The radon problem is especially serious for underground miners as the gas accumulates in the tunnels.

Less well known aspects of radon gas are its use as a means of measuring earthquake activity, and its influence on weather. Emissions of radon gas from natural cave systems are used to measure and predict earthquakes and earth tremors. When ground motion deep below the surface crushes rock, radon is released. Even the slightest tremor shows up as a burst of radon gas. The mouths of cave systems all over the planet are wired to "sniff" radon and warn of earthquake activity. An unexpected discovery during this research was that natural radon emissions occur all the time and play a role in the ionization of gas and water molecules in the air, creating lightning and affecting weather. The huge quantity of radon coming from uranium mine and mill wastes may have an effect on weather patterns.[34]

Heavy Metals And Process Chemicals

Radioactive materials are not the only hazardous component of uranium mine and mill wastes. Also of concern are heavy metals, which are a potential problem with any type of mining. In 1978 the AECB Advisory Panel on Tailings wrote:

Tailings also contain concentrations which vary from site to site of heavy metals such as lead, zinc, manganese, cadmium, and arsenic whose rates of release to the environment must also be controlled. It must be remembered however, that elements such as these do not decrease in toxicity with time since there is no decay process, they simply last forever.[35]

Heavy metal poisoning is usually noticeable long before any effects of radioactivity. It is the heavy metals and process chemicals that are the primary reason plants and fish die downstream from uranium mines. Huge quantities of process chemicals are used in the milling process, then dumped into the environment. For example, according to Eldorado Nuclear's 1984 annual report to the AECB, the Rabbit Lake mill used about 200 tonnes of concentrated sulphuric acid per day.[36] Some of the other chemicals used include: ammonia gas, hydrochloric acid, kerosene, and hydrogen peroxide.[37]

The main Rabbit Lake mill tailings pond and Wollaston Lake

Photo: Christine Rognerud

Large Volume

Large volumes of waste are produced in the uranium milling process over a short period of time. Hundreds of tonnes of waste are produced for every tonne of yellowcake. Official reports minimize the volume by focusing attention on the solid component only, or "tailings." But liquid wastes have a greater impact on the surrounding environment than solid wastes as they can carry contamination great distances via streams, rivers and lakes. The highly toxic liquid waste from a uranium mill is usually more than twice as big in volume as the solid waste. What is more, liquid wastes are continually being added to by surface and ground water seepage through waste areas.

According to Eldorado's 1984 annual report to the AECB, the Rabbit Lake mill discharges 7.7 million litres of waste water per day to the tailings disposal area.[38] This contaminated water flows into Wollaston Lake at Hidden Bay via two settling ponds. The maximum holding capacity of the two ponds together is reached after only 16 days of mill operation.[39] In addition, contaminated water flows into Pow Bay via a series of drainage ditches that divert surface water runoff around the Rabbit Lake pit.[40] There is thus a steady flow of waste water into Wollaston Lake at two points.

The large volume of just the solid radioactive wastes produced by a uranium mill is hard to comprehend. Table 1 shows the quantity at Saskatchewan uranium mines.[41] The 4 million tonnes of solid radioactive mill wastes produced by the Rabbit Lake mine alone is enough to cover almost knee deep a two lane highway 800 kilometres long.[42] That is about the distance by road all the way from the Rabbit Lake mine to Saskatoon. Another way of expressing the large volume of waste created at uranium mines is by considering the amount produced to make an amount of reactor fuel, as illustrated in Figure 4.

In January 1987 production of solid uranium mill wastes in Canada reached at least 130 million tonnes – about 110 in Ontario and 20 in Saskatchewan. This amount represents a volume easily capable of covering a two lane highway a metre deep all the way from Vancouver to Halifax, coast to coast.[43] Added to this already huge quantity is the gigantic amount of rock not containing enough uranium to put through the mill, and all the the rock that had to be removed in order to reach the uranium.

Solid Radioactive Uranium Mill Wastes In Northern Saskatchewan – Present and Prospective

Mine	Quantity (metric tonnes)	Years Of Operation
Uranium City Area:		
Beaverlodge	6,000,000	1952-82
Gunnar	5,500,000	1955-64
Lorado	360,000	1957-60
Rabbit Lake	4,000,000	1975-85
Collin's Bay	1,930,000	1985-91
Cluff Lake		
Phase I	84,000	1981-84
Phase II	2,700,000	1984-95
Key Lake	4,500,000	1982-2000
Total	25,074,000	

Figure 4

Uranium Mill Wastes Produced To Make Fuel Used By A Nuclear Reactor

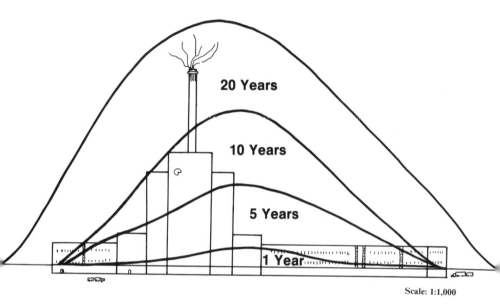

20 Years

10 Years

5 Years

1 Year

Scale: 1:1,000

The figure illustrates the quantity of uranium mill wastes (solid and liquid) produced in order to make the nuclear fuel burned during operation of the Forsmark 3 nuclear reactor. Shown are the quantity of uranium mill wastes produced after 1, 5, 10, and 20 years of reactor operation, in relation to the size of the reactor buildings (which cover an area of 40,000 square metres, or about 10 soccer fields). The reactor, in Sweden, is 1050MW.

The assumptions made are: production of 1 kg of yellowcake creates 1,000 kg of solid and 2,000 kg of liquid mill waste; 1 m^3 of waste weighs 1.6 tonnes; 1 kg of yellowcake is taken to be equal to 1 kg of natural uranium; to make 1 kg of reactor fuel requires 3.7 kg natural uranium; and the reactor burns 30 tonnes of fuel per year. Thus, about 330,000 tonnes, or 210,000 m^3, of uranium mill waste are produced to run the reactor for 1 year. The content of radioactivity in the waste is about 40,000 Bq/kg.

Source: Mats Törnquist, Söderboda 3601, 74071 Öregrund. December 2, 1986.

One of the three Forsmark nuclear reactors in Sweden, with their scrap metal garbage dump in the foreground.

Accumulation Of Radioactivity In Plants And Animals

The April 26, 1986 nuclear reactor meltdown at Chernobyl in the Soviet Union has shown once again that the Earth is one vast, interrelated ecosystem. In a matter of weeks, the radioisotopes from a few tonnes of uranium fuel were carried by the winds and taken up into the cellular tissue of plants and animals around the world.

Even with the knowledge gained from bomb fallout and the Chernobyl and other accidents, scientists do not have a detailed understanding of the biological accumulation of radioisotopes. Generally, research is limited to quantifying bioaccumulation in the aquatic environment without examining the impact of that bioaccumulation. The concentration of some radioisotopes increases at higher levels of the food chain, while others decrease. A complexity of factors are involved. However, it is well accepted that the uptake of radioisotopes does occur, and that the rate of uptake is highly dependent on the species, food chain, ecosystem, and isotope.[44][45] In any case, bioaccumulation mobilizes radionuclides within ecosystems, making them more easily available to consumers, including human beings.

Many edible plants in northern Saskatchewan rapidly accumulate radioactivity. Some of them are blueberry (*Vaccinium myrtiloides*) and labrador tea (*Ledum groenbandicum*);[46] examples from more southern latitudes include: alfalfa, clover, peas and beans,[47] and seaweed.[48]

In the Canadian context, only a few studies examine accumulation of radioactivity by mammals. This area warrants more attention as there have been two reports of a cow moose carrying a two-headed fetus being shot near Wollaston Lake. Further, Wollaston residents have often killed moose from the Rabbit Lake mine area, and people have reported seeing moose drinking from the tailings ponds.

In February, 1978 a photo was taken of Wollaston resident Gabriel St. Pierre holding a two headed moose fetus taken from a cow moose he shot near Wollaston Lake. The photo was used in a poster made in 1981 by the Group For Survival, a Saskatoon native rights group. As such mutations often occur in livestock

110

Wollaston resident Gabriel St. Pierre holding a two-headed moose fetus taken from a cow moose he shot near Wollaston Lake in February, 1978.

on southern farms, a common response to the photo is that it was just a coincidence that the mutation occurred near a uranium mine. In answer to this, it certainly is no coincidence that the major food source of the moose, aquatic plants, greatly concentrates radioactivity. This is confirmed by the Dubyna Lake plant samples noted below. Further, in the pre-uranium mining

111

óral history of the Wollaston native people, such mutations are completely unknown. In addition, similar livestock mutations on southern farms are certainly a defect.

Research from other parts of the world helps to understand the Canadian situation. For example, a 1971 study in the Soviet Union found the effect on small mammals living in areas with high uranium-238 and radium-226 concentrations is greater incidence of sterility. It was also found that gamma radiation reduced bird populations by reducing the number of hatching eggs.[49]

The variability in concentration of radioactivity by species and isotope is illustrated well by sampling carried out at the Rabbit Lake, Dubyna, and Beaverlodge mine sites.

Beaverlodge Mine

The downstream area from Eldorado Nuclear Ltd.'s Beaverlodge mine and mill at Uranium City has been studied extensively. In a 1977 sample from about 1 km downstream from the mill, approximately 25% of lake chub (*Couesius plumbeus*) caught had eye deformities. Some of the fish had one or both pupils very small, and other fish had lens cataracts in one or both eyes. The laboratory report analyzing the fish reads:

There was no evidence of infection or parasitic encystment within the eye. Cataracts may result from genetic makeup, nutritional deficiency, environmental effects, or a combination of the three. Certain factors such as high levels of radiation, parasitic infection, or the presence of specific chemicals can contribute to cataract formation.[50]

The longnose sucker (*Catostomus catostomus*) at right was caught in the summer of 1982 downstream from the Beaverlodge mine at Uranium City. The fish is totally blind. The eyes have no pupils at all. The mouth of the sucker is especially adapted for eating off the bottom, where it spends most of its time. Since radioactive particles are heavier than water they quickly settle out and accumulate in the bottom sediment of streams and lakes. Thus, bottom feeding fish such as suckers suffer more from the effects of radiation than other species.

Fish from the downstream area of the mine and mill caught in July 1979 contained 10 to 100 times as much radioactivity as

112

Blind fish caught in 1982 downstream from the Beaverlodge mine.

fish from an uncontaminated lake.[51] The researcher conducting the analysis concluded:

The main transfer pathway of radionuclides in Beaverlodge Lake appears to be via contact with the sediments, either directly or through food organisms. White suckers, feeding on benthic invertebrates had the highest radionuclide content; lake whitefish, feeding on a combination of chironomids[52], lake chub, stickleback and benthic algae had moderate levels; while lake trout, feeding mainly on cisco, a planktivorous species, had low levels.

The differences in radionuclide content among tissues reflect chemical and physiological factors. Radium-226 and lead-210 replace calcium in the bone matrix resulting in high concentrations of these radionuclides in bone. Uranium also collects primarily in bone although the critical organ is the kidney.

The high skin levels for all three radionuclides may be due to particulate matter adhering to the outer surface rather than actual incorporation into skin tissue.

Skin is high in calcium; therefore selective tissue incorporation may also contribute to high radium-226 and lead-210 levels in this tissue.[53]

113

It is important to note that Beaverlodge Lake was closed to commercial fishing in the mid-1980's.

Dubyna Mine

The Dubyna mine is owned by Eldorado Nuclear Ltd. and is located 12 km northeast of Uranium City. A bioaccumulation study was done there before mining began, when the site had only been cleared of trees and drilled in a grid pattern. It is unknown if data is available from during and after mining, and before exploration.

The sampling results show that due to the intense drilling, levels of radioisotopes in plants and fish are thousands of times greater than levels in the surrounding water, and that the degree of uptake is isotope and species specific. For example, of the three aquatic plants waterlily (*Nuphar variegatum*), millfoil (*Myriophyllum alternifloram*), and sedge (*Carex aquatilis*), mill-foil concentrates uranium the greatest (at 14,000 times) while waterlily concentrates greater amounts of radium (at 11,000 times), and sedge the greatest amount of lead-210 (at 13,000 times).

Regarding fish, both northern pike (*Esox lucius*) and lake trout (*Salvelinus namayush*) accumulate radioisotopes more in the bone (up to 11,000 times) than in the flesh (up to 6,500 times). The degree of concentration though, is species specific. Lake trout were found to have greater levels of uranium, thorium and lead-210, but northern pike had the greatest level of radium.[54]

Rabbit Lake Area

A 1978 Canadian Ministry of Environment study of the area immediately downstream of the Rabbit Lake mill, called Effluent Creek, concluded that:

The biological community of Effluent Creek was undergoing considerable stress, as evidenced by a lack of zooplankton, and low diversity of phytoplankton... [a] preliminary study showed high

Animal tracks (1985) on the radioactive waste left by the Lorado uranium mill near Uranium City, which operated from 1957–60.

ammonia concentrations throughout Effluent Creek and an almost complete absence of benthic invertebrates in bottom samples collected along the entire length of the creek.[55]

As distance increases from the source of contamination the effect on plants and animals is no longer so obvious. Effluent Creek flows into Hidden Bay which opens up into Wollaston Lake. The 1978 study noted:

There were consistent indications from the various biological communities (zooplankton, phytoplankton and benthic invertebrates) that there was at least localized impact in Hidden Bay in the vicinity of Effluent Creek mouth.[56]

In terms of impact on fish, the study documents that toxicity tests of the Rabbit Lake waste discharge "on several occasions found the tailings effluent acutely lethal to rainbow trout (*Salmo gairdneri*)."[57] Laboratory tests from March 1977 to January 1979 putting rainbow trout in precipitation pond effluent, found that all the fish died in 96 hours, even when the effluent concentration was only 10%. In July 1978 tests were conducted putting sucker fry (*Catostomus sp.*) collected from Collin's Bay into plastic containers submerged for 56 hours in Effluent Creek, the inlet to Horseshoe Lake, and precipitation pond effluent. The water was found to be acutely lethal in all but the Effluent Creek sample. It is noted that the water contained toxic levels of ammonia.[58]

Vegetation in the Rabbit Lake area has also been analyzed. In 1983 a researcher from the Department of Biology at the University of Saskatchewan determined quantities of lead-210 and polonium-210 in vegetation at two sites in the Rabbit Lake area, Collin's Creek and Hidden Bay, and for comparison purposes, two sites near the Churchill River: Birch Hill and Otter Rapids.[59] The Rabbit Lake sites showed significantly greater accumulation in four of the ten species analyzed: blueberry (*Vaccinium myrtiloides*), labrador tea (*Ledum groenbandicum*), green alder (*Alnus crispa*), and black spruce (*Picea mariana*). Collin's Creek was found to be a "hot spot" for all species except dry-ground cranberry (*Vaccinium vitis-idaea*). A different study in another area looking at uranium levels in trees found the greatest amount in the growing tips of twigs, followed by bark, leaves and wood.[60]

The 1983 Rabbit Lake study concluded:
– Plant groups accumulate the two radionuclides in the follow-

116

ing order, from greatest to least: lichen, moss, shrubs, and trees. Lichens and mosses accumulate five and ten times the levels of shrubs and trees respectively.
- Among the shrubs, blueberry accumulates the greatest amount of radionuclides.
- Primary uptake in the shrubs is through the root systems, whereas uptake in the moss and lichens is from their surfaces following deposition from the air. The source is therefore local for the shrubs and trees but more widespread for the moss and lichens.[61]

Radioactivity Transfer To People

Radioactivity in the environment eventually finds its way to people, particularly because they are at the end of a food chain. Numerous studies have shown that radioactivity from bomb fallout and reactor accidents reaches people through the food chain. A biological pathway to people that has been confirmed through scientific study is the lichen–reindeer/caribou–human food chain.

Studies in northern Scandinavia and Canada on the lichen–reindeer/caribou–human biological pathway found that people consuming the reindeer or caribou that ate the lichen ended up with several times the normal level of radioactivity in their bodies. The reindeer of Scandinavia and caribou of Canada are members of the same species. Certain lichen species are the main food of both animals.

The concentration of radioactivity in the atmosphere decreased after the above ground nuclear test ban treaty in 1963. However, due to previous accumulation in plants and animals, the risk to humans continued. From 1965 to 1969, annual surveys were carried out to determine the human body content of cesium-137 in Inuit from 25 northern Canadian communities. The results showed levels 20 to 100 times those of people living in the south. In some cases the level was so high that it even exceeded the safe limit suggested by the conservative International Commission on Radiological Protection.[62] Further, Inuit could accumulate sufficient lead-210 to double the total skeletal radiation dose over southerners.[63] In Finland, the exposed people were found to have eight times the normal level of polonium-210 in their blood.[64]

Lichen accumulate greater amounts of radioactive fallout, be it from bombs or uranium mines, than most other plants. The

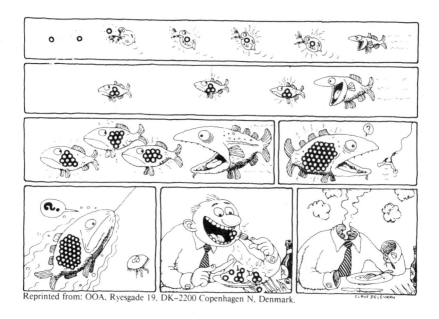

Reprinted from: OOA, Ryesgade 19, DK–2200 Copenhagen N, Denmark.

reason is threefold. First, it is because they obtain all of their nutrients from the air, such as from rain and dust that falls on them. Second, they have a much greater effective surface area than most plants. Third, as they are so slow growing, they are exposed over a longer period of time. The higher than usual accumulation levels in lichen occur for non-radioactive trace elements as well.

Though humans can receive an elevated radiation dose via the lichen-reindeer/caribou food chain, the accumulation levels of most radionuclides in this pathway progressively decline. For example, lead-210 and radium-226 levels in wolves have been found to be approximately one-tenth those in caribou. The behavior of radionuclides in the aquatic environment is similar.[65]

No Solution In Sight

No-one has been able to figure out how to stop long-term pollution from uranium mill wastes. There are no reports from anywhere in the world of a uranium tailings area being successfully put into a state so that the spread of contamination is stopped. The AECB Advisory Panel On Tailings stated bluntly in 1978:

The existing waste management system is inadequate as a long-term solution. It is undesirable and unrealistic to rely for the integrity of any method of waste management no matter how efficient, on human intervention over thousands of years.[66]

In agreement, a report prepared by the Canadian Ministry of Environment in 1980 stated:

It must be recognized that environmentally acceptable methods of long-term mill tailings disposal which would not be dependent on continued human management have yet to be developed.[67]

A detailed international study by the International Atomic Energy Agency that ended in 1986 came to the same conclusion.[68]

Meanwhile, contamination is reaching far into ecological cycles, and because the uranium industry in Canada is vigorously expanding, the volume of wastes is rapidly accumulating. The AECB is not unaware of the problems that will be encountered. In 1978, the AECB Advisory Panel On Tailings wrote:

The anticipated expansion of the Canadian uranium mining and milling industry emphasizes the need to find an acceptable long-term tailings management system as soon as possible. The longer it is necessary to use waste management methods which in the long-term are inadequate, the larger the volume of wastes that will be generated and the more difficult it will be to apply retroactive, remedial procedures.[69]

No "long-term tailings management system" has yet been found. Spills continue and attempts at limiting contamination by revegetation have done little to change contaminant levels, as discussed below.

"Spills"

It is common practice for regulatory agencies to monitor so called "spills." However, in terms of impact on the local ecology, all the wastes, 100% of them, are "spills" into the environment. Dams and other barriers only serve to limit the real extent of the wastes, and even that they do not do efficiently. Liquid and solid wastes regularly break through their retention barriers.

The consequences of a waste release depend on its size and the time before remedial action is taken, for which records are seldom available. The range of remedial measures include: re-routing streams, adding treatment facilities downstream of the point of release, removal of contaminated material, rebuilding dykes and dams, and, of course, doing nothing at all.

Many spills have occurred at the Rabbit Lake mine and mill. On January 15, 1980 Mr. J. A. Keily, then Vice President of Production and Engineering for Gulf Minerals Rabbit Lake mine and mill, made a presentation at the British Columbia Royal Commission of Inquiry into Uranium Mining (B.C. RCUM). The following exchange took place between him and Kris Bogild, lawyer for the West Coast Environmental Law Association:

Mr. Keily: Well, spills are inevitable in any process you have and as a matter of cleanliness, the yellowcake packaging area is vac-uumed or washed down after every packaging run.
Kris Bogild: Thank you. How many spills have you had since start up?
Mr. Keily: If we are talking of spills of yellowcake, I would say I know of no more than one.
Kris Bogild: And what other types of spills have you had?
Mr. Keily: We have spills of slurry, we have spills of ground ore, we have spills of water.
Kris Bogild: Would the slurry, ground ore, and water all contain radioactive material?
Mr. Keily: Yes.
Kris Bogild: And how many of these spills have occurred since start up?
Mr. Keily: Probably too numerous to count.[70]

In 1982 about 44,000 litres of mill water spilled into Banana Lake (immediately downstream from the mill). People in Woll-aston were not informed by the mining company. A teacher who was in Prince Albert at the time heard about the spill on the radio news and informed the community. The cause of the spill was human error. Workers changing a pipe leading into the tail-ings pond had shut off the flow, but someone at the other end, not knowing the pipe was disconnected, turned the flow back on.[71]

Revegetation

The waste piles in Canada cannot be revegetated because of an overabundance of ground and surface water, and in some cases acidity. Still, Eldorado Nuclear claims that the Rabbit Lake tailings area will be turned into "a nice moose pasture."

In any event, it is important to realize that plant growth on top of a tailings area does not mean the spread of contamination is stopped. Limited plant growth has been achieved with massive fertilizer application and natural plants have regrown along the edges. But plant growth can actually increase the quantity of radon gas escaping from the wastes. This is because radium travels up through the roots and is distributed in the leaves. Thus the surface area available for radon release is greatly increased.

In addition, root penetration allows water to seep through the protective soil cover and into the tailings, allowing ground water to be polluted. As well, the plants themselves become contaminated through uptake of toxic materials, which pose a danger to any animals eating them.

The AECB Advisory Panel on Tailings accepts that revegetation is not a solution to the waste problem. In 1978 they wrote:

A number of successful experimental plots of vegetation have been grown on tailings. None of the sites, however, is regarded by the regulatory authorities as having been decommissioned, meaning that none of the sites could be left by the operator with no provision for further care and maintenance.[72]

Similar conclusions have been reached in the U.S. A study by the Los Alamos Scientific laboratory to investigate revegetation of 21 uranium mill waste piles in the western U.S. concluded that revegetation would not control release of radioactivity and heavy metals.[73]

Remedial Action

A number of actions can be taken to try and minimize the spread of contamination from uranium mine wastes in northern Saskatchewan. The most obvious is stopping to produce more waste. In any case, some actions that could be taken where con-

tamination already exists are:
- put up fencing to keep large mammals away, and "scare crows" to stop birds from landing,
- install fencing in contaminated streams to stop fish from entering,
- post signs warning people not to drink water or harvest food,
- improve radium and thorium removal equipment,
- measure radioactivity transfer to mammals, including people,
- begin a "local monitoring society" with the responsibility of passing knowledge to future generations on how to deal with the wastes.

Precautions Against Radioactivity

Precautions can be taken to decrease one's accumulation of radioactivity when eating contaminated food. However, when radioactive particles are breathed in, exposure by ingestion may represent only a small portion of total exposure to radioactivity.

A high calcium diet can reduce accumulation of some radionuclides. Since radium-226, lead-210, strontium-90, and cesium-137 are analogous to calcium, they concentrate in bone.[74] If the maximum amount of calcium needed by the body is available, there is less chance the radionuclides will take its place and a greater chance they will be discarded in body wastes.

The same effect can take place in animals and plants when adequate sodium and potassium (analogues to cesium-137) are available to them. This principle has been applied in Scandinavia in an attempt to lessen the effects of radioactive fallout from the Chernobyl accident. Some people added potassium to their vegetable gardens and placed sodium and potassium rich lickstones in the forest for wild animals.

The use of calcium, sodium, and potassium supplements in uranium mining areas needs to be investigated. Radium-226 and lead-210 are two of the primary pollutants from uranium mines, and can cause serious damage because of their long biological half-lives (10–12 years, as compared to uranium-238's 100 days).

Another way of reducing the uptake of radioactivity from contaminated food is by selective eating. That is, by limiting consumption of the most contaminated species, and parts of species. In such a strategy fish bones and skin would not be eaten, and eating of bottom feeding fish would be minimized.

122

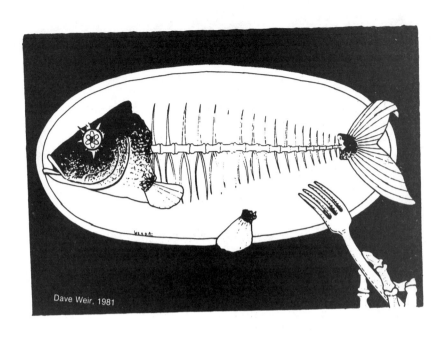

Dave Weir, 1981

The Illegal Operation Of
The Rabbit Lake Mine

Before uranium mines can operate they need a license from the Atomic Energy Control Board (AECB).[75] Some aspects of the license requirements are:

- Surface water quality limits are set for only eight pollutants. arsenic, copper, lead, nickel, zinc, total suspended matter, pH, and radium-226, though waste water contains many more heavy metals and radioisotopes.
- Regular monitoring is required at a place referred to as the "final point of control." This point is usually several kilometres downstream from where the wastes first leave the outlet pipe. The area between the mine and mill and the "final point of control" is a "sacrifice zone" where no regulations apply.
- The ground water seepage problem is not addressed. Eldorado intends to dump the mill wastes from the Collin's Bay operation in the mined-out Rabbit Lake pit. It is impossible to prevent radioactive contaminants from entering into the pit area ground water, which flows into Wollaston Lake.

123

The Rabbit Lake pit, 550 metres wide and 150 metres deep.

Even with these inadequacies, it is not easy for uranium mining companies to comply with the license requirements. In fact, uranium mines can not operate without exceeding pollution limits set out in the license. This is documented in the mining companies' own annual license compliance reports submitted to the AECB. Simply comparing waste discharge monitoring data in the annual reports to license limits shows the limits are often exceeded. Even the amount of pollution discharge allowed is dangerous to life forms. For many heavy metals and radioisotopes the quantity allowed by the AECB license is above that suitable for human drinking water and for aquatic life. Thus, a Canadian government license to operate a uranium mine is a license to kill aquatic life and poison drinking water.

In 1978 J. H. Jennekens, Director-General of the AECB Operations Directorate sent the annual operating license for the Rabbit Lake mine and mill to R. N. Taylor, president of Gulf Minerals Canada Ltd., with a covering letter that read:

During the 1977 operating year, certain effluent guidelines and regulations with respect to levels and concentration of deleterious

124

substances were exceeded on occasion. As discussed, attention must be given to these areas so Gulf Minerals Limited will meet the criteria in the future.[76]

All the criteria have never been met. Not surprisingly, in May, 1985 when Eldorado was taken to court in La Ronge, Saskatchewan over their pollution of the area downstream from the Rabbit Lake mill, the Provincial government "entered a stay of proceedings." This means that the case could not be heard. A "stay" completely bypasses the judge and the government is under no obligation to give any reason for its action.

Following are examples of lead, arsenic, uranium, and radium downstream from the Rabbit Lake mill. When considering any individual pollutant it is important to keep in mind cumulative effects with other toxic materials. A known effect of one pollutant would probably occur at lower concentrations when other toxicants are present.[77]

Lead

The AECB license requires that the monthly average content of water pollution samples must not contain more than 200 ppb lead. However, ENL's 1983 license compliance report records an average of 250 ppb for the whole year, exceeding the license requirement by 25%. Further, the limit set for any single sample of lead (400 ppb) is recorded as being fully two times as high (800 ppb) as is allowed by the license. But what does this mean for aquatic life and human beings?

The AECB discharge limit for lead (200 ppb) is four times greater than the quantity Health and Welfare (H & W) Canada allows in drinking water (50 ppb) and six and two-thirds times greater than what the United States Environmental Protection Agency (US EPA) says is dangerous to aquatic life (30 ppb), and twice as high as what is known to kill minnows (100 ppb).

Arsenic

With regards to arsenic H & W Canada allows a "maximum acceptable concentration" of 50 ppb in drinking water, but the AECB license allows ten times that amount (500 ppb).

125

Uranium

For uranium, H & W Canada specifies 20 ppb as the "maximum acceptable concentration" for drinking water. For the whole year of 1984, according to Eldorado Nuclear's own report, the average level was 135 times greater (2,700 ppb) at the airport road drainage (a "final point of control"), which drains into Pow Bay. The other "final point of control," called the precipitation pond outflow, is immediately downstream of the tailings ponds. There, the average uranium content in February 1984 was 1,500 ppb. This is not surprising considering that the tailings dumped during the whole of 1984 contained uranium equivalent to about 100 tonnes of yellowcake.[78]

Radium

The AECB license limit (monthly average) for discharge of radium is .37 Bq/1., while the Saskatchewan government's Surface Water Quality Objective is set at the more stringent level of .11 Bq/1. H & W Canada's "maximum acceptable concentration" for drinking water is 1 Bq/1., and a target concentration has been set of .1 Bq/1. All these levels have been exceeded regularly.

In 1978 the AECB wrote to the operators of the Rabbit Lake mine requesting a reduction in radium levels at the airport road drainage (one of the "final points of control") to below 10 pCi/l (.37 Bq/l).[79] The last available data reviewed, from 1984, reported radium levels still above this level.

Saskmont Engineering, a consulting company working for Gulf Minerals Canada Ltd., completed an investigation of radium levels in the Rabbit Lake drainage system at the airport road drainage from January 1977 to September 1981.[80] According to the data they compiled, the .37 Bq/l AECB license limit is exceeded 42 out of 57 months examined; and the H & W Canada 1 Bq/l "maximum acceptable concentration" for drinking water is exceeded in 15 months.

In a letter dated August 18, 1982 from Gulf to Saskatchewan Environment it is stated:

Radium at the Precipitation Pond discharge (weir no. 2) exceeded the maximum allowable monthly average in February, April and June.

Among the reasons given for the high levels were power

126

failures and extreme cold.[81]

Further, the 1982 Gulf compliance report reads:

Radium-226 sampled on a weekly composite basis, exceeded the specified limit 17 times... The yearly mean radium-226 value for 1982 is .406 Bq/l which is higher than the 1981 level of .16 Bq/l.[82]

The average quantity discharged for October 1984 was .44 Bq/l and for April 1984 even higher at .55 Bq/l.

Position Of Canadian Authorities And Eldorado Nuclear Ltd.

Similar positions on the issue of uranium mine pollution are taken by the province of Saskatchewan, the federal government of Canada and Eldorado Nuclear Ltd. All admit the streams flowing from the Rabbit Lake mine and mill area into Wollaston Lake are not as clean as they were in their natural state. But they say once any pollution present mixes in the open lake it is no longer dangerous. This philosophy is known as "the solution to pollution is dilution." Eldorado's annual reports show otherwise.

At Hidden Bay, a point downstream from where the liquid waste from the Rabbit Lake uranium mill is dumped, ENL recorded an average of 60 ppb lead in 1983 and 40 ppb in 1984. Both these levels are above the US EPA danger level for aquatic life (30 ppb), and close to the H & W Canada maximum acceptable for drinking water (50 ppb). It is important to note here that these ENL numbers are "averages", meaning that some of the samples had to be higher. Further, there was a regular flow of thousands of litres of polluted water a day from the Rabbit Lake uranium mill. Localized impacts from such large quantities of polluted water cannot be avoided.

Federal Government

John Witteman, Director of the Saskatchewan District Office

127

of the Canadian Ministry of Environment, stressed at a town meeting in Wollaston Lake on April 17, 1985 that the mines are being operated in a "safe" manner. He also stated mine wastes have never killed fish, contradicting the published findings of his own Ministry.[83] One is left to decide if John Witteman is uninformed, intentionally misleads, has a definition of "safe" that allows a high degree of pollution, or some combination of the three.

Nonetheless, Mr. Witteman got a shocked response from the local residents when he explained that in a government fish test it is OK if up to 50% of the fish die. He was describing the "LC50" test, where the lethal concentration of a substance in water is found that will kill 50% of a population of an organism over a certain period of time. A clearer example of the cultural differences between the government and indigenous people of the north could not be found.

With regard to the AECB, Bernie Zgola, of the Uranium Mine Waste Management Division, stated at the April 17, 1985 community meeting in Wollaston:

Because of the fact that it is uranium, the mining of uranium is controlled much, much more strictly and conducted in a much, much safer manner than the mining of other minerals. Does that offer a better degree of safety? I would say yes.... Our responsibility is to ensure that the mining of uranium in any area of Canada is done in a safe fashion.

In a further effort to convince the Wollaston people that uranium mining is safe, Larry G. Chamney of the AECB Waste Management Division put together a short report titled, "Summary of Water Quality and Fish Analysis In Wollaston Lake Near The Rabbit Lake/Collin's Bay B-Zone Facility," dated May 28, 1985. The report used water quality data from Hidden, Collin's and Pow Bays averaged over 10 years to conclude no hazard is presented. This is misleading as: danger levels were exceeded for several parameters in individual years; cumulative, long-term effects were not considered, which are often very serious; and the area between the mine and Wollaston Lake was not examined.

The report by Mr. Chamney goes to the extreme of making the false statement that,

It is intended that no contaminated water will be released to Collin's Bay as a result of the B-zone facility.[84]

128

The steel dyke separating Collin's Bay from the B-zone pit.

The dyke separating Collin's Bay from the B-zone pit will be destroyed after the planned 6 years of mining. When the dyke is destroyed, the loose rock in the pit will be a source of heavy metal and radioactive contamination. In addition, all life on the bottom of the Bay in the immediate area of the dyke will be completely destroyed by suffocation from sand and gravel.

Provincial Government

The Mines Pollution Control Branch of the Saskatchewan Ministry of Environment published a report in February, 1985 that reviewed the past and future environmental impact of Eldorado's Rabbit Lake/Collin's Bay operation. The report concludes:

...the levels of uranium and radium are not of concern. No significant impact has been observed nor is any expected.[85]

What is more, the report gives the impression that any radio-activity reaching Wollaston Lake from Collin's Bay is due to "natural processes."[86]

129

Eldorado Nuclear Limited

One of the ways Eldorado responded to the controversy around its activities at Collin's Bay in the spring of 1985 was by increasing hiring efforts in Wollaston, without success. They also arranged a series of three special public relations mine tours for Wollaston residents. The tours took place on March 26 and April 3 and 11, 1985. Though Eldorado had taken over the Rabbit Lake and Collin's Bay B-zone operations three years earlier, it was the first time they invited Wollaston residents to take a look.

The tours included return air transport from Wollaston to the mine site, lunch and supper, and visits to the mines and mill. As well, a film about the wonderful uses of uranium was shown in a small theater and followed by a question and answer period. The visitors were given hardhats, cotton overcoats and rubber boots.

On the first tour, all the teachers in town were invited. They were not impressed. According to one teacher the Eldorado representative lied a number of times. He said yellowcake was not radioactive and remarked more than once how the tailings area would be turned into a nice moose pasture, something that is technically impossible. Another teacher noted that stress was put on trying to convince them the operation was safe, that the Eldorado staff were very pleasant and had obviously rehearsed their performance.

The second Eldorado mine tour, on April 3, 1985, was for "distinguished residents of Wollaston Lake," as the invitations read. People invited included the Local Advisory Council (LAC), the Band Council, Father Megret, the nurses and a couple of elders.

The experience was especially significant for elder Louis Benonie who had never been to the mine site before. He used to trap where the mine is now and was given "compensation" of $12,000 plus $1,000 a year for life by Gulf Minerals, who owned the mine at the time. Near the end of the tour Mr. Benonie spoke up in front of everyone. He said,

I came to hear what you have to say, but I want to have my say too. I would never have given my trapline to you if I would have known what you were going to do. I almost cried when I saw what you've done to my trapline. This is my land. You people don't belong here. I thought you were going to use uranium for something good, like money, like gold, I didn't know you were going to use it to kill each other. If I would have known that I never would

130

have let the mine go ahead.

The mining companies make a great effort to hide the fact that they employ such a small number of native people. After the mine tour LAC staff person Terry Daniels, who had also taken part in a mine tour conducted by Gulf Minerals in about 1977, said,

Then, they had an Indian man standing in each room. When one woman asked one of these men what he did, he said, 'I was just asked to come in here and stand around.' At least this time they never went that far.

The third and last in the series of three PR mine tours for people from Wollaston Lake was on April 11, 1985. This tour was specially arranged for the adult education class. Eldorado did not allow the three pregnant woman in the class to take part.

Eldorado Nuclear hides the pollution problems from the mine as openly as the AECB and Ministry of Environment. Despite clear documentation in their own reports that the AECB license requirements for operation of the Rabbit Lake mill have not been met, Eldorado states otherwise. A May 1985 letter given to Eldorado employees from Wollaston Lake reads:

Eldorado continues to meet or outperform all of the objectives for employee, public and environmental protection, as required in its AECB license.[87]

At the same time, industry officials are not unaware that uranium mines pollute water. This was made clear at a 1981 public hearing that took place in La Ronge. There, John Keily, the Gulf vice-president in charge of the Rabbit Lake mine, was asked if he would drink from a jar containing a water sample taken downstream from the Rabbit Lake mine. His answer, recorded in the meeting transcript, was, "I won't recommend that anyone drink that water" (which flows into Wollaston Lake).[88]

Summary Of Conclusions

- Canada is the western world's greatest producer and exporter of uranium. Without purchases and investment from abroad coupled with extensive government subsidization, the Canadian uranium industry cannot survive.
- It is not possible to separate the civil and military nuclear industries. Some of the same processing facilities are used in the production of nuclear fuel for civil nuclear reactors as in the production of the two critical explosive components of nuclear bombs, plutonium and uranium. Canadian uranium mines directly and indirectly supply nuclear weapons programs.
- Uranium mining has a catastrophic effect on the area immediately surrounding the mines and contaminates the downstream region.
- Uranium mining rapidly produces large volumes of liquid and solid waste, which remain hazardous for thousands of years.
- Uranium mine and mill wastes spread primarily through water transport into complex ecological cycles. The radioactivity accumulates in plants and animals downstream to levels thousands of times the surrounding water concentration. This contamination can eventually find its way to people.
- Accidental release of uranium mine waste from their retention barriers is common.
- Attempts at limiting the spread of contamination from uranium mine waste areas by revegetation have not been successful.
- Lead, radium, arsenic and uranium levels downstream from the Rabbit Lake mine and mill exceed the quantity allowed by the AECB license. For these same parameters, the quantity allowed by the AECB license is above that suitable for human drinking water and for aquatic life. Thus, a Canadian government license to operate a uranium mine is a license to kill aquatic life and poison drinking water.
- At the present time there is no known method of stopping the spread of contamination from uranium mine wastes.
- The Canadian government agencies responsible for protection of the environment at both the provincial and federal levels continue to allow exploration and new mines to proceed. At the same time, the authorities and Eldorado Nuclear Ltd. often try to hide the serious pollution problems caused by uranium mining, and clearly state the pollution is "not significant."

CHAPTER 3
THE
RESISTANCE

Save
Wollaston Lake

T. Roberton '85

Leave Uranium
in the Ground

"Lack of understanding is obviously a major problem facing the nuclear industry and the most powerful ally of the anti-nuclear movement."

"Without understanding, there is little resistance to fear. And there is no doubt about the social and political force of that weapon: examples are thrown at us daily in North America, and this is certainly true in Canada. Lack of knowledge and understanding are played upon. Latent fears are thereby activated, at least to the point of raising questions in the minds of our citizens. Reacting to this new public interest in nuclear matters, our elected representatives are eager to be seen asking questions, rather than providing answers. And questions beg more questions.

We are rapidly reaching the point where the industry, the anti-nuclear movement, the public and their elected representatives are not debating nuclear energy. We are all running a hard race on a circular track with no finishing line in sight. The first to drop from exhaustion loses."

Nicholas M. Ediger, Chairman, President and Chief Executive Officer, Eldorado Nuclear Ltd.[1]

The page at left is from a leaflet made by members of the Regina Group For A Non-nuclear Society. The graphic was donated by Teri Robertson of Vancouver.

History
Of Opposition

People from Wollaston Lake have consistently spoken out against uranium mining ever since their first contact with the Rabbit Lake mine project. Opposition was voiced at meetings held in Wollaston with government and mining company officials in 1972 and 1977, a public hearing held in La Ronge on July 28 and 29, 1981 dealing with the Collin's Bay B-zone project, and numerous meetings in the north throughout 1984 and 1985. On several occasions during 1984 and 1985 Wollaston residents attended anti-nuclear meetings in the south, and outside supporters attended community meetings in Wollaston. The minutes of a meeting on April 30, 1985 with the Inter-church Uranium Committee, a coalition of the major churches in Saskatchewan that was formed to oppose uranium mining, are included here.

1977

Two Wollaston residents interviewed in June 1984 recalled the 1977 meeting.

Joseph Besskaystare:
In 1977 I was Chief of the Lac La Hache Band of Wollaston Lake. There was a meeting here about the Rabbit Lake mine. About seven people came in and the priest was there as a translator. Those people were talking about employing all the young people at the mine and paying royalties to the people. They were going to

White fish caught in Wollaston Lake. A major reason for the opposition to uranium mining is concern about contamination of the fish in Wollaston Lake.

Mary Ann Kkailther smoking meat and fish.

make a big store here, but I told them "No" to the mine. I told them no to the mining because of what it might do to the lake. I told them you guys can move around but us living here we don't want to move just because of the mine. In about 35 years you people will be finished mining. All the workers will be gone but we will still be here. After the water is contaminated what are we going to live on? If I said okay to the royalties the money will stop coming when the mining is finished but the water will still be contaminated.

Melanie St. Pierre:

In 1977 my brother was a Chief and there was a big meeting at the school. There were a lot of people from the government. A lot of people from here came. Those people that came asked if they could open up the mine and everybody was against it. At that time Joseph Besskaystare didn't want the mine to be opened. Even if we are going to get that royalty money, we're not going to give permission to open up that mine.

We were told at that meeting that at that time Germany and Russia were making military weapons with uranium but that Canada and the United States didn't have any.

Even if we get so much money in each family, people won't take it because we're thinking of our kids in the future. Maybe they make millions of dollars from the mine at Rabbit Lake, but the

Wollaston fishermen delivering their catch to the government operated Freshwater Fish Marketing Corporation at barge landing.

people at Wollaston don't need it because they're not getting anything. We have no use for the mine.

We mothers have a lot of young children. The greatest concern of the mothers today is what the children are going to live on if the water, land and animals are destroyed.[1]

1981

The public hearing on the Collin's Bay B-zone project held in La Ronge July 28 and 29, 1981 was the last of four Saskatchewan government hosted meetings. The government stated that the purpose of the meetings was to "receive the opinions and suggestions of Saskatchewan people." However, right from the beginning, the government viewed mining in general as acceptable. The Saskatchewan Ministry of Environment's role was more to contribute engineering studies on technical aspects of waste management rather than present evidence on environmental impacts. At that time, the Collin's Bay B-zone deposit was owned 50% by SMDC and 50% by Gulf Minerals and Noranda.

Gulf showed their worst side at the meeting. They paid all the expenses for a handful of native employees (from communities

north of Wollaston) to attend the meeting, including charter of a small aircraft. Gulf even went so far as to write some of their native employees' statements. This disgusted the appointed Chipewyan-English translator, Marie Rose Yooya, who resigned on the second day.

On July 28, 1981, the first day of the hearing, Wollaston resident Terry Daniels is quoted in the transcript as saying:

There was a remark made earlier that all the northern communities were in favor of uranium development. I'm afraid that Wollaston is always being left out. I want to say now that Wollaston doesn't go along with what these people are saying.

The next day Emil Hansen, then Chairman of the Wollaston LAC read a one page written submission at the hearing. It is reprinted in its entirety below.[2]

Presentation By the Wollaston LAC to the Collin's Bay B-zone public hearing in La Ronge, July 29, 1981

We as a community have made it very clear to Gulf Minerals in the very recent past that we are very much opposed to the proposed expansion of uranium mining in the Collin's Bay area on Wollaston Lake. The exploration and mining activities that have been in progress for the past 10 to 15 years have provided very little, if any, benefits to the area residents.

Ten years ago when the Rabbit Lake mine was being planned, there were no public hearings to our knowledge so no one really knows how much the people of the area opposed that project.

We are still original and traditional natural resource users even though it was stated by one of the pro-development supporters yesterday that we no longer have to depend upon the land for a livelihood. The land is still very much a part of us and we would certainly like to see it protected for us and our children and their children.

These opinions are shared by every member of our community, but have never been broadly stated as we feel that like many other times, it would only fall upon deaf ears.

If the people who presented briefs yesterday supporting the Collin's Bay project really believe what they said then any statements they may make in the future regarding pro-uranium activities certainly do not represent the feelings of the people of our area. Let me make it clear that native employees of Gulf Minerals and the traditional resource users do not share the same views.

An environmental impact study should be carried out. The traplines and the fishing areas that will be affected not only by the Collin's Bay B-zone but also by the several other huge projects that will be proposed not only by Gulf but by other multinational corporations should be stated and documented. We believe that this procedure is also hopeless but at least the feelings and the statements of our area will be documented.

We are opposed to Gulf Minerals' dirty tactics during this public review by using northern native people as puppets to justify their trespassing and the destruction of a traditional way of life on this land. In closing we say that we know that the public hearing process is only a formality and anything we say at this public review will not have any degree of influence on the final decision on whether or not the mine should go ahead.

1984

In mid-July 1984 an open letter was sent from the Wollaston Lake Lac La Hache Band and Local Advisory Council appealing for support to stop uranium mining. The beginning and end of the three page letter is included below. This letter was circulated widely and resulted in a steady flow of support mail to the Lac La Hache Band. Attached to the letter was a petition against uranium mining. Thousands of signatures were received at the Band office by the summer of 1985.

April 30, 1985 - Meeting With The Inter-church Uranium Committee (ICUC)

At the invitation of the Wollaston people the following church leaders attended a community meeting in the Wollaston Band Hall, April 30, 1985: Bishop James Mahoney, Roman Catholic Bishop, Saskatoon; Sister Margaret Ordway, Saskatoon; Edgar Epp, Executive Director of the Mennonite Central Committee of Saskatchewan; Joan McMurtry, Incumbent President of the Saskatchewan Conference of the United Church; Tom Powell, President of the Saskatchewan Conference of the United Church; John Cliner, Senior Professor at the Saskatoon Lutheran Seminary; and Peter Prebble, ICUC Saskatoon. Also in attendance was George Smith, Mayor of Pinehouse and President of the Saskatchewan Association of Northern Local Governments (SANLG). Most of George Smith's statement has been included in the chapter called "The People." After the meeting in the Band Hall most of the group went on a tour of the Rabbit Lake mine site.

Following are excerpts from the community meeting minutes. Marie Rose Yooya, a professional Chipewyan-English translator and District Representative for the Athabascan Bands, was present for the occasion.

Chief Hector Kkailther:

Talks were held with people in town before this meeting so some people are prepared. Please make your comments brief and easy to translate. The floor is open to hear people's views. A meeting with the provincial government is being organized and comments here will be used for that meeting.

Elder Louis Benonie:

When the mine was started a meeting was held in the community. The present mine site is on my traditional trapline. At the meeting people opposed building of the mine. At a meeting two years ago the present mining at Collin's Bay and plans to mine underwater in the future were opposed. When I was on a tour of the mine I saw what it looks like. Even the land looks pitiful. I wonder why these things happen the way they do. Every time there is a meeting the people have opposed mining. It doesn't seem that anything is being done. Everyone wants to sit down and have a meeting but nothing is being done. We've heard a lot about uran-

Photo: Flynn Swan

ium development and the effect on people, and we are concerned about the future. If the land and water is contaminated what will the children live on?

Thomas Sha'oulle, Band Councilor:

The concerns that are being raised about Rabbit Lake are coming from all of us. We oppose the mining because of the problems. When we had the meeting with the government officials they said they won't stop mining until all the fish are destroyed. What is the sense of stopping the mining then? You can see that the older people here have always lived off the land. What will the children live on if they don't have fish? The people in the community all have the same concerns. We rely on the fish and caribou. Now we have to go quite a ways to get at the caribou; they used to come closer.

Another tactic the mining company is using now that we are

protesting is they're just hiring left and right, even young boys. We the people of Wollaston Lake oppose uranium development.

Elder Bart Dzeylion:

As far as we are concerned we are opposed to uranium mining. We are all of the same mind on that. Every time the government comes we tell them the same thing. Every time there is a meeting we are asked for our opinions and we give our opinions but it seems that is where it stops. Some may write it down but they, the mines, go ahead anyway.

Not only us here are going to be affected by this mining, especially if they start mining underneath the lake. There are two major tributaries coming out of Wollaston Lake. One goes into Alberta and the other down to Reindeer Lake. These areas will be affected. The white people that have been exploring told us this is a very good body of water, it is so clear. But what they were doing

145

was searching for uranium and they didn't tell us.

This is half the community here. If everyone in the community came in here, including all the children, no one would be able to move. We don't want our children to be put in a pitiful situation. If you people that came here are going to help us we would be very pleased for that. We by ourselves cannot do it. If you were able to help us accomplish these things we would be very happy. Us people here, what we think and feel is that money is nothing. If you lose money you can survive but if you lose one human life that is what we are trying to protect, human life.

Elder Odest Dzeylion:

When they first started talking about uranium development the people opposed it. My reason for making these comments is that my traditional trapline is right beside where the mine is. I used to be able to live there. There was moose and fur bearing animals but now there are no animals there. Before I could always draw a living from the land. Now I have no source of livelihood. Never once did they come to me and say we will compensate you. As far as I'm concerned I'd like those people to leave, to stop uranium mining and quit destroying the land. Now we have a block of land that we can use and we can't set a trap outside it. If they're going to be doing that to us we don't want them here. What are my children and their children going to live on?

George Smith:

Last month at our Association meeting I met Hector and agreed to support you in your uranium fight. I'm speaking for 21 communities in the north. I think you people are really concerned. I've never seen so many people come to a meeting.

Elder Tony Dzeylion:

The things that are being talked about here were talked about at several meetings in Saskatoon that Mary Ann Kkailther and I went to. People laughed and said, "Why are you going down there?" Now one of the results of going to those meetings is that the church leaders came here.

And I'd like to say a few things about Louis getting $1,000 a year. The head boss of the mine kicked me off three times. One person is getting money, but there are many people living here and we only get $1,000!

Marie Rose Yooya:

One of the tactics the company uses is to use a little bit of

146

Elder Tony Dzeylion.

compensation to say, "We are compensating." But it's like Tony Dzeylion says, what about the whole community?

Sister Margaret Ordway:

Last time I was here as a messenger to set up a meeting with the Cabinet and to see if you agreed to the way we drew up the plans for you to present your petitions. The Christian leaders wanted to come here before they go with you, and for you, to the government. I can speak for them and say we hear you in your concern for the land and the future. It is something strengthening for us. There remains something important for us to do today: the plans for the meeting and the press conference in conjunction with the meeting.[3] I hope Bart will never be able to say again that nothing has come out of a meeting something will happen.

Pierre Besskaystare:

When the mine first opened they hired about seven of us saying we would be given jobs, but they just gave us shovels and asked us to do the dirty work. And when someone got hurt they didn't get paid even if they were hurt while doing the work. Commercial fishermen were told that if there was any sign of uranium in the fish they were harvesting, the mine would close. It sure doesn't look like that now. For us it is very hard to do things around here, but the mining companies go in and do what they please. It is not only the feeling of the commercial fishermen, also the women and children want the mining stopped.

Chief Hector Kkailther:

These people that are here visiting and listening, it would be good if they can help us, but if it doesn't work I hope they can come back.

The mine has been in operation for about ten years now and we have been rarely informed, just minimally, now it is just lately that we are finding out. That is why we are doing what we are doing. It's not only the present mining development at Rabbit Lake but the future Collin's Bay B-zone and there are others, about 11 in the area. If you were going to put all those together, it is a very large development, a big concern. It's not just our concern. Our visitors have been listening to us. I'm sure they'll deal with it. Thank you for this meeting.

Wollaston commercial fishermen.

148

Fresh Fish

I'm going to a place
where there's fresh fish to eat
year-round any time you want
and as much as you like.
There are no shopping centres in Wollaston.
I'd rather have fresh fish
than a shopping centre anytime.
I'm looking out the window of the plane
and seeing lakes dotting the landscape
all the way to the horizon
and they all probably have fish in them.

Do you know what happens
when you eat fresh fish regularly?
Pretty soon you can't tell the difference
between yourself and the lake.
You can't draw a line between yourself
and the water, the fish and the forest.

You walk through the forest
and you're like that fish
swimming through the water
that is in your blood.
The fish is the land and you are the fish
so you are the land.
Your connection to the Earth is clear.
There's no cement
no asphalt separating you.

Do you want fresh fish?
Stop uranium mining!

**– Flying Swan, on the way to
Wollaston Lake community, March, 1985.**

150

The Gathering And Blockade June 9–17, 1985

NORTHERN SURVIVAL GATHERING

work shops

Northern Peoples Concerns
International Nu–Killers
Appropriate Developments
Blockade Preparations

june 9 – 13 , 85

june 14, 85
ROLLING BLOCKADE
WOLLASTON LAKE
ROAD
SASKATCHEWAN

Honor Earth Life People

STOP URANIUM MINING

"If the water is contaminated and not fit to drink and the fish are not fit to eat, WHAT ARE THE CHILDREN GOING TO LIVE ON ?"

Poster made by the Group For Survival, a Saskatoon based native rights group.

A road sign in northern Saskatchewan.

The Collin's Bay Action Group

The Collin's Bay Action Group (CBAG) is a coalition of native rights and anti-uranium groups. It was formed in December 1984 for the purpose of coordinating the June 1985 blockade and preceding gathering. It was understood from the beginning that Wollaston people would lead the protest, and that no violence, alcohol, or drugs would be allowed.

Wollaston people set the dates June 9–13 for the gathering and June 14 for the beginning of the blockade, with no planned end. Support groups in several cities began having organizing meetings and the planned events started to be discussed at regular community meetings in Wollaston Lake.

The following five pages show some examples of preparation work done by CBAG. Following these are some letters and resolutions of support written before the blockade took place.

152

The Collin's Bay Action Group

We are a group of concerned individuals formed to:

- support the people of northern Saskatchewan in their fight for self-determination and against uranium mining, and

- strengthen the international native rights anti-nuclear movement by spreading information about the devastation of the people and land that has already taken place in northern Saskatchewan.

We recognize our relationship to all the animal and plant life of the land and water, the natural elements, and human life on Earth.

We view the uranium open pit mines or any such disturbance of the land as an act of violence. We oppose the destruction of our land and our way of life. We condemn the mining ventures of all multinational corporations for the rape of our Sacred Mother Earth.

Our goal is survival, for ourselves, our families, and yet unborn. We pray for the survival of the new generations.

OPEN INVITATION TO NORTHERN GATHERING AND BLOCKADE

We ask that those who come to the survival gathering and rolling blockade be sensitive and committed to working with different cultures, and respect the land we will use.

The survival gathering is a place of work, not a festival. We ask that people come prepared and as self-sufficient as possible. A women's space will be provided. Warm clothes, sleeping bags, and tents are needed. The camping area will be designated. We ask that people not wander around or camp outside the designated area.

You are the guests of the Wollaston people. We ask you to act in solidarity with our efforts. Many people have worked hard to form a united front and provide a life giving future for our children.

Those who exploit us and the Earth have played too long on our differences. Our appearance and behavior will be closely watched and judged by others who have yet to come forward and those who have just begun to. For this reason, as well as to keep the survival gathering and blockade peaceful and respectful, we have established the following guidelines.

GUIDELINES FOR PARTICIPATION
IN THE NORTHERN GATHERING AND BLOCKADE

NO ALCOHOL OR DRUGS

NO DESTRUCTION OF LAND OR PROPERTY

RESPECT FOR ELDERS, CHILDREN,
DIFFERENT CULTURES AND ONE ANOTHER

ALL OUTSIDE LITERATURE MUST BE APPROVED

Security/peacekeepers will be on site to aid you, these
people will be working around the clock to ensure a safe,
peaceful gathering and blockade. Please respect their
advise as they have been specially instructed in dealing
with problems.

These guidelines might be difficult for some to agree
with, but the values of many who live here may be
different from those people who live outside this area.

WHEREAS, such sacred elements to our survival as food and
water are becoming increasingly controlled by profit
minded multi-national corporations;

WHEREAS, multi-national corporations intend to further
exploit Indian treaty lands by extracting oil, gas, coal,
uranium and other minerals; and that uranium is being
mined in northern Saskatchewan in complete disregard of
native land claims;

WHEREAS, radioactive contamination from the use of
nuclear fuels poses a dire threat to all life on our
sacred Mother Earth; and

WHEREAS, political and social freedom is threatened by
concentration of economic power in the hands of the
directors of the multi-national corporations and
representatives of the Tri-lateral Commission;

WHEREAS, the escalating nuclear arms race poses an
immediate threat of universal annihilation;

BE IT RESOLVED, that the health and basic needs of the
people of Wollaston Lake and northern Saskatchewan take
priority over the rights of the multi-national
corporations and government; and

BE IT RESOLVED, that an indefinite moratorium be declared
by all progressive nations on the mining, milling,
processing, and fissioning of uranium.

H. E. L. P.   Honor Earth Life People

154

Why Stop Uranium Mining?

*Uranium mining is taking place in complete disregard of native land claims and aboriginal rights.

*Uranium is the material that makes nuclear bombs go boom and that is turned into deadly plutonium in nuclear reactors.

*Uranium mining rapidly produces large volumes of liquid and solid waste, which remain hazardous forever.

*Uranium mining has a catastrophic effect on the immediate surrounding environment and contaminates the downstream area.

*There is no existing method of stopping the spread of contamination from uranium mine wastes.

*Accidental release of uranium mine wastes from their retention barriers is common.

*Radioactivity released from uranium mines accumulates in plants and animals downstream to levels thousands of times the surrounding water concentration. This contamination can eventually find its way to people.

*Uranium miners can die of cancer and contract serious lung diseases as a direct result of working in uranium mines.

Sponsored by the Collins Bay Action Group Wollaston Lake, Sask. S0J 3L0 Tel. 633-2003

The above advertisement was placed in "The Northerner" on April 24, 1985. It is the newspaper published in La Ronge, with a circulation of about 2,000.

Adele Ratt's home in La Ronge, June 1985.

The Northerner La Ronge, Saskatchewan June 5, 1985

Letters to the Editor cont.

Dear Editor,
Re: Why civil disobedience at Wollaston.

It is important for people to understand why the residents of Wollaston Lake are opposed to uranium mining. It is very simple. Uranium mining threatens the very lifeblood of the people and the land. It threatens the lifeblood of all forms of life - WATER. We cannot live without water. Wollaston Lake is being threatened with extinction by the mining of uranium. The contamination of water by uranium mining has created many 'dead' lakes and rivers all over the world.

One example of the devastation that has been done to people is the contamination of the Pine Ridge Indian Reservation in South Dakota. In 1980, the Women of all Red Nations (WARN) conducted a health study on the Pine Ridge Indian Reservation (a part of the Oglala Sioux tribe). They found, as a direct result of uranium mining; unusually high rates of spontaneous abortion (6½ times the national average), cancer and genetic defects and breathing complications in newborns.

Although cancer is historically not common among Indians, the current incidence among Lakota (Sioux) women of breast and uterine cancer, as well as leukemia and sterility, has risen to epidemic proportions.

Two towns were to receive new wells, because of radioactive contamination, but drilling attempts had been unsuccessful in finding clean water. The Oglala Sioux tribe has declared a state of emergency on the reservation.

The WARN report also found that in this same area, on surrounding ranches, the rate of stillborn or deformed calves has skyrocketed. WARN said, "To contaminate Indian water is an act of war more subtle than military aggression, yet no more deadly."

Everywhere in the world

156

Enlargement of a button made by The Group For Survival.

where there is uranium mining, water is contaminated. This damage is irreparable and irreversible. It is important to understand that water is a non-renewable resource. Once it is gone - it is gone forever. Uranium mining uses vast amounts of water throughout the entire process.

If the water in Wollaston Lake is contaminated - how will the people survive? Already in the north, thousands of gallons of highly radioactive uranium mine wastes have directly been dumped into lakes and rivers. This does not include the most recent Key Lake and Rabbit Lake tailings spills.

In the Uranium City area, there are reported 'dead lakes', yet no fences or warning signs are put up to keep people out or to keep them from drinking the water. Birds and animals can't read signs, even if they were up. What are the government and corporations doing about that?

Essentially, what the government and corporations are guilty of is 'premeditated mass murder', by allowing the deliberate contamination of water. There is a law in Canada which says that if you do not report a crime - you are committing a crime. By not trying to stop this spread of contamination of northern lakes and rivers. In my opinion, you are aiding and abetting mass murder! Thousands of people in the north directly depend on northern lakes and rivers for their lives. By allowing pollution, I feel the government and corporations have a licence to kill and the people of Saskatchewan are allowing this insanity to occur. We want to declare the entire north of Saskatchewan as in a state of emergency right now!

The people of Wollaston Lake have been protesting the mining of uranium since 1972. Now, they are forced into civil disobedience in a desperate attempt to save their land, their water and their way of life.

How far the government is willing to go will be determined at the June 14 blockage of the Wollaston Lake Road. We will see, then, how far the people of Saskatchewan will allow them to go.

Support the people's actions!

Adele Ratt

157

Some Letters And Resolutions Against Uranium Mining

Mennonite
Central
Committee
Saskatchewan

2206 Speers Avenue
Saskatoon
Saskatchewan
S7L 5X7

Telephone
(306) 665-2555

MCC

*A Christian
resource
for meeting
human need*

ACTION 8:   RESOLUTION ON URANIUM MINING - MCCS Annual Meeting, November, 1982

*Whereas the Scriptures clearly call us to be peacemakers, and
Whereas delegates to the Annual Meeting of MCC (Sask.) in 1977
resolved to call for a five-year moratorium on uranium development
in the province until the moral question, the environment question
and the safety question be further explored, and
Whereas these questions have not been given 'adequate explan-
ations and answers" during those five years, and
Whereas uranium mined in Saskatchewan is known to have been used
in the manufacture of nuclear weapons, and
Whereas there is to date no evidence to suggest that Saskatchewan
uranium will not continue to be used in nuclear weaponry.*

*Be it resolved that MCC (Sask.) call on the government of
Saskatchewan to place an immediate moratorium on uranium mining until
the moral, environmental and safety questions are adequately dealt
with, and until such time it can be guranteed that no uranium mined
in Saskatchewan will or can be used in the manufacture and further
proliferation of nuclear weaponry.* Carried.

The Northern Village of Pinehouse
GEN. DEL. PINEHOUSE, SASK.
SOJ 2B0

Telephone 884-2030

Chief Hector Kkailther,
Lac La Hache Band Administration,
Wollaston Lake, Sask., SOJ 3CO.   January 21/85.

Dear Hector:

At our council meeting on December 13/84 we passed a motion
supporting 100% the Lac La Hache Band and Wollaston LAC in their
fight against the Collins Bay uranium mine.

Putting a uranium mine on the bottom of a lake is one of the
stupidest things I've ever heard of. Uranium mining is no good.
Fishing is much more important. Our planning study shows that for
Pinehouse what we get in one year from commercial and domestic
fishing is worth over 3 times what we get into town in wages from
the Key Lake mine. But the worse thing is that the uranium pollution might poison your fish for decades. It's an awful bad
deal for you guys and we're happy you're fighting it.

We went to Regina in October to talk against using chemical
poisons on the bush. Tony Dzeylion was there to tell the federal
government that you're against uranium mining. He talked really
strong and clear to those guys. We know you guys are against
them mining from the bottom of the same lake that you get your
fish and water from. Anybody who isn't against such a stupid
and dangerous idea is either being paid by the uranium industry
or just plain crazy.

                    Sincerely,

                    George Smith, Mayor.

c.c. band councillors Martin Josie, Jimmy Kkaikka, Rosalie Tsannie,
     LAC councillors Flora Natomagan, Gabriel Benonie, secretary-
     treasurer Emil Hansen, and mayor Jonas Hansen.

EXECUTIVE RESOLUTIONS

#1 (Submitted by Sask. Div. Executive Board)

WHEREAS several proposed uranium developments around Wollaston Lake
by companies like Eldorado Nuclear, Hooker Chemical and
Esso pose a pollution threat to the northern environment and
could cause serious long term damage to the fishery resource
and way of life of the Lac La Hache Indian Band; and

WHEREAS uranium sales by the Saskatchewan Government to buyers such
as the military dictatorship of South Korea and COGEMA (the
French agency involved in atomic weapons testing in the
Pacific Ocean) are contributing to the spread of the atomic
bomb around the world; and

WHEREAS uranium mine workers face cancer rates more than double the
national average and a government in this province that has
weakened occupational health and safety provisions that had
been in place; and

WHEREAS no country in the world has found a way to safely dispose of
the highly cancer causing and radioactive waste that all
uranium becomes after being burned in a nuclear reactor.

BE IT RESOLVED that the Saskatchewan Division of C.U.P.E. expresses
its full support for and solidarity with the Lac La Hache
Indian Band in their attempt to stop uranium mining around
Wollaston Lake and calls on the provincial and federal
governments to immediately stop the Collins Bay B Zone
uranium mine (currently under construction in a bay of
Wollaston Lake).

BE IT FURTHER RESOLVED that the Saskatchewan Division of C.U.P.E.
reaffirm its opposition to uranium mining in Saskatchewan
and calls upon all political parties in this province to
support the phase out of existing uranium mines.

BE IT FURTHER RESOLVED alternative employment and full transfer of
benefits must be assured for all employees effected.

BE IT FURTHER RESOLVED that a letter of solidarity be sent to Chief
Hector Kkhailther of the Lac La Hache Indian Band and that
copies of this resolution be forwarded to Prime Minister
Mulroney, Federal NDP Environment Critic Bill Blaikie,
Federal Liberal Disarmament Critic Lloyd Axworthy, Saskatchewan
Premier Grant Devine, NDP Opposition Leader Allan Blakeney,
Liberal Leader Ralph Goodale and the appropriate members of
the media.

*Resolution made at the Canadian Union of Public Employees
(CUPE) annual convention May 24-26, 1985. (CUPE, c/o Sask-
atchewan Federation of Labor, 2709-12 Ave., Rm. 103, Regina,
Saskatchewan S4T 1J3)*

160

Inter-Church Uranium Committee
Box 7724 Saskatoon Canada S7K 4R4 (306) 934-3030

Eldor Management and Staff,
#2115 - 11th Street, West,   Monday, June 3, 1985
Saskatoon, Saskatchewan

Dear Members of Eldor Management and Staff,

For at least 8 years now, many Saskatchewan churches represented on our Committee have been urging your company and other
companies involved in uranium mining to halt uranium development
in Saskatchewan. We have accumulated a large amount of evidence
pointing to the fact that:

- Uranium mining is directly linked to nuclear weaponry,
  despite good intentions to sell it only for peaceful
  purposes. Uranium exports contribute to weapons proliferation.

- No country on earth has solved the problem of what to do
  with the highly radioactive wastes that all uranium will
  become after its use in nuclear power plants.

- Uranium mining is a serious long term threat to the
  Saskatchewan northern environment.

- Uranium mine workers have significantly increased incidences
  of lung cancer and other types of cancers.

We have particularly deep concerns about Eldor's decision to
proceed with uranium mining along the edge of Wollaston Lake. As
a federal government agency, it is important that you be sensitive
to public concerns and particularly the concerns of people who will
be directly effected by your developments.

The community of Wollaston Lake is very much opposed to uranium
mining around their lake. It legitimately fears the pollution of the
lake, and with several proposed developments right along the edge
of the lake, we believe this is a legitimate fear, despite the good
intentions of Eldor to mine as safely as possible.

It is simply not appropriate to locate a uranium mine within little
more than one hundred yards of one of the richest commercial fishery
lakes in Saskatchewan and a lake which an Indian people is so highly
dependent on for their livelihood. Eldor will mine uranium for only
a few years and then be gone. But the people of Wollaston must live
with the consequences you create for hundreds of thousands of years
to come.

This weekend, the Saskatchewan Conference of the United Church
of Canada, of which I am a member, reaffirmed its call for a halt to
uranium mining in Saskatchewan and expressed its deep desire that
Eldor stop the proposed Collins Bay B Zone project at the edge of

Wollaston Lake. For the sake of the people of Wollaston Lake
and for the sake of the Saskatchewan environment, we plead with
you to abandon the proposed Collins Bay B Zone uranium development.

To bring our concerns before the public and to the Government
of Canada to which you as a company are directly responsible,
we will be conducting a very short noon hour rally calling for a
halt to uranium mining around Wollaston Lake.

Thank you for considering our concerns.

            Sincerely,

            Alan Richards
            Rev. Alan Richards
            on behalf of the Inter-Church Uranium Committee

The DIOCESE OF SASKATCHEWAN
OF THE ANGLICAN CHURCH OF CANADA

BISHOP: The Right Reverend H. V. R. Short, D.D.

June 6, 1985

Chief Hector Kkailther
Lac la Hache Band
General Delivery
Wollaston Lake, Sask.
S0J 3C0

My dear Chief:

At a recent meeting of the Executive Committee of the Anglican Diocese of Saskatchewan, a resolution was unanimously passed requesting me to convey to you the support of the Anglican Diocese of Saskatchewan in your protest against any further development of uranium mining at the present time in the Wollaston Lake area.

This I am very happy to do, and I wish to add to it my own personal support of the stand you are taking. It seems to me that there are three questions that must be addressed before the natural resources of the country are developed, and they are especially in connection with uranium mining; the possible dangers attending radioactive tailings on mine sites; the possible use of uranium for the manufacture of atomic warheads; and the question of land entitlement for aboriginal people.

As we pray for justice your own particular concerns will be in our minds.

With every blessing.

Yours sincerely,

+Vicars Short.

Bishop of Saskatchewan

kw

NORTHLANDS INDIAN BAND

GENERAL DELIVERY
LAC BROCHET, MANITOBA
R0B 2E0

Lac La Hache Band
Wollaston,Sask. and L.A.C

Dear Chief,

This letter is to inform the Northlands Indian Band at Lac Brochet
Manitoba are in support of the movement your Band is making in the apposing
of the opening of the uranium mine in Wollaston Lake - Collin Bay B. Zone
uranium ore body.

This was decided at the general meeting in Lac Brochet in regards
to the support needed from your Band.

We realized also that it affect us in the future for we are situtated
at Lac Brochet where the Coccrance River flows from Wollaston Lake and in on to
Reindeer Lake.

Hope this letter also acts as a guideline of support for your future
endeavors with your respective Government of Saskatchewan in your future negotiations .

Hope we have been of same help on behalf of the Northlands Band.

Sincerly,
for Indian Self
Government.
Chief Jerome Denechezhe
& Councillors

163

Collins Bay Action Group,
Wollaston Lake

Dear friends

I'm speaking for people from the
Ottawa community, we hope that things
are going well with you in northern
Saskatchewan. We'd like to hear alot more
about what's going on (on a more
regular basis if possible) as not much
gets out in the (state-controlled) news.
We'll try to put some pressure on M.P.'s
and such in this area, but there are lots
of things going on this summer.

Included with this letter is a check
addressed to the Collins Bay Action
Group. The amount is $626, money raised
from a benefit concert evening we had
here in Ottawa. It was a great evening
The concert was put on by us (RAR) and
another group, Youth Cultural Promotions.
Co-sponsorship came from Pollution Probe,
the Ontario Public Interest Research Group
(OPIRG - Ottawa chapters), Skyline (a local
cable TV station), and two radio stations,
CFUO (University of Ottawa) and CKCU (Carleton
University. The cable station and radio station
did interviews with various people about
the show and about the protest.

Over two hundred people came out to the
show, and literature was passed out
at the door. Hopefully most people
understood the meaning of the gig.
The bands were great, with a variety
of different music. All local, and
playing for free (with expenses such
as gas) the bands were: The Trapt,
The Calamity Janes, Honest Injun and
Fluid Waffle.

I know that this money will be needed.
Please, if anything is needed in Ottawa
(information, accomadations, etc) write
us. Remember, if possible, to keep us
up to date on what's happening.
With the strength and spirit of the
people, we can overcome the struggle.
In strength, mind and spirit, we will
be with you. Take care, and don't give
up. Wich I was up there with you,
in struggle.

In Peace ~ Joe Chang

Rock
Against · Ottawa
Racism

Movement Against Uranium Mining
PO Box K133 Haymarket 2000 Phone 212 4538

May 9 1985

Collin's Bay Action Group
Box 3183
Vancouver BC VGB 3X6
Canada

Dear friends

Our very best wishes for your rolling blockade of the Rabbit Lake and
Collin's Bay uranium mines beginning mid-June. We hope your inspiring
action will not only raise public awareness of the devastating problems
that result from uranium mining, but also be a source of encouragement
for the whole anti-nuclear movement in challenging this insidious
industry.

Please keep in touch through the World Information Service on Energy
network so we can report the blockade to the Australian public. Our
thoughts are with you.

Yours in solidarity and struggle

Steve Broadbent.

Steve Broadbent

PS I've enclosed a copy of Uranium Mining in Australia and a copy of
our last newsletter, which has updates on the uranium mining industry
in Australia.

OOA Organisationen til Oplysning om Atomkraft
Ryesgade 19
DK - 2200 København N
☎01-35 55 07 & 01-35 48 07

Copenhagen, June 6th, 1985

TO THE INDIAN PEOPLE OF WOLLASTON LAKE AND THE PARTICIPANTS

IN THE BLOCKADE OF THE COLLINS BAY MINE, SASKATCHEWAN, CANADA

The Danish antinuclear movement OOA (The Organisation for
Information on Nuclear Energy) sends its warmest regards and
its full moral support to the planned Survival Gathering and
the non-violent blockade of the Collins Bay mine at Wollaston
Lake, Saskatchewan, Canada.

We are shocked by the degradation of the natural environment and
all the encroachments against the Indian people, that the mining
companies have caused during their hunt for uranium in Saskatchewan. We consider any mining of uranium completely unacceptable,
whether the purposé is the production of nuclear weapons or running nuclear power plants, and we regard the international struggle
against uranium mining as an important and necessary means to stop
the whole military-industrial nuclear complex.

In the hope that we can share with you some of our optimism, we
can inform you, that we now - after 11 years of stubborn public
resistance - have succeded in getting the Danish Parliament to
abolish all plans for nuclear power. 15 sites which were reserved
for nuclear power plants years ago, have now been released for
other uses. This is being celebrated by people living near these
sites, who are now in the process of raising monuments to mark
the occasion.

Furthermore, both Danish authorities and the Greenlandic home
rule have decided to give up the plans for uranium mining in
Southern Greenland. This is a great victory for the people of
Greenland.

...2

166

These experiences tell us, that it surely is worthwhile to protest and fight back, even though the powers you oppose are great. We know that your counterpart is more implacable and more cynical than ours, but we also know that you are a strong and determined people. We hope with all of our heart, that you will endure. Because there is no alternative.

For economical reasons we are unfortunatly not able to come to Wollaston Lake and participate in your blockade. But on June 10th we and six other Danish organisations will carry through a protest action at the Canadian Embassy in Copenhagen to show our solidarity with your struggle. We will inform the Danish and the Canadian press about this action.

We enclose 150 Canadian dollars as a contribution to your work.

Best regards

for OOA

Elisabet Stadler

Bjarne Hellskov Nielsen

Lena Warrer

Jørgen Steen Nielsen

Marianne Juhl-Nyholm

Margrete Aronsø

Rolf Nielsson

Stichting Nanai

nederlandse actiegroep noord-amerikaanse indianen

Centraal kontaktadres: Kamgras 23
3068 CB Rotterdam
telefoon 010·209844
postgiro 3520900

Amsterdam 19-6-1985
to:Lac La Hache Band
   administration,
   Wollaston Lake
   Saskatchewan SOJ 3CO
   Canada

Dear people,

Enclosed a number of petitions concerning the
uranium mining in your area.
We hope you will have a lot of succes with your
actions.
In Holland the government wants to build more nuclear
power stations.The majority of the Dutch people is
against the plan.But the government doesn't want to
listen.As it is the same in Saskatchewan:your struggle
is our struggle!
Sincerely,

Fred Muller

Nanai,region Amsterdam
Fred Muller
Wethouder Serrurierstraat 66
1107 CH Amsterdam
Holland

168

Photo: Raimo Long

The road north from La Ronge to Wollaston is long and narrow with barely room for two trucks to pass (bottom). Mickey's Camp at McLennan Lake, 130 km north of La Ronge, is one of the few gas stations along the way. It has a wall filled with photos of truck accidents (above).

Photo: Lillebror

The Gathering,
June 9–13, 1985

June 9, 1985:

Small groups of supporters from across Canada arrived at the Umperville River campground after an 800 kilometre, 12 hour drive from Saskatoon. The campground is only about 10 kilo-

A point in the road about halfway between La Ronge and Wollaston Lake.

metres south of the Rabbit Lake open pit mine, and 40 kilometres west across the lake from the village of Wollaston Lake.

In preparation, just a few weeks previously, Eldorado built a new gate across the road with fence extending about 30 metres on each side. It was erected about two kilometres from the mill buildings, cleverly placed just around a corner where the mill buildings were out of view and keeping open an access to one of the waste areas. On the gate, which does not mark any legal boundary, a sign read "No Trespassing: Trespassers Will Be Prosecuted," in English, French, Cree, and Chipewyan syllabics.

June 10, 1985:

About 30 supporters held meetings in the campground, and named themselves the "Wollaston Lake Support Group." A special session on sensitivity to cultural differences was held. The group was invited into the community of Wollaston Lake to meet with the people. In the evening they were greeted by a welcome dance performed by Wollaston's own young rock music group, the Northern Eagles.

June 11 and 12, 1985:

The meetings between outside supporters and Wollaston Lake

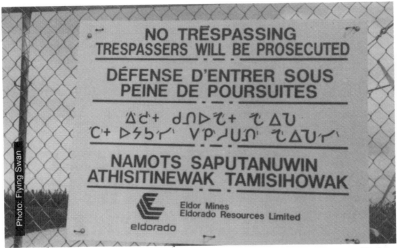

When a group of Wollaston Elders saw the Chipewyan syllabics on the mine gate sign (middle row on the sign above) they expressed their astonishment and wondered who had done the translation. They said the words meant that harm would come to those that went on the other side.

community members filled the Band Hall to capacity. Strong statements were made from young and old alike to stop uranium mining. Each evening a dance followed the long meetings.

On June 12 a community meeting began at 8 in the evening and ended at 1:30 in the morning. The Northern Eagles then played to a lively group that danced until the music stopped at 5 in the morning.

The meetings were held in the traditional Dene manner. There was no formal agenda. People spoke one after the other continuously, each one speaking until they were finished, without being interrupted. The group of 150 or more listened intently throughout. Almost everything was translated from Chipewyan to

171

English and vice versa. Most of the meeting time was devoted to statements by northern people. In addition, a group of southern supporters performed a play depicting Eldorado's schemes to mine uranium and information was provided on other land based struggles.

All of the following was stated at the June 11 and 12 meetings.

Statements Made At Meetings In The Wollaston Lake Band Hall – June 11 and 12, 1985

Bengy Denechezhe (Lac Brochet),
opening prayer for the meeting:

Brothers and sisters we are gathered here on a concern to all of us. We are not rich but we are happy. We were raised poor but we managed through hard times and good times with warm loving hearts and the beauty of outdoor life. When we look back it's a crying shame compared to now-a-days. But one thing for sure, we will never forget our traditional way of living. Yesterday will never come back to us. We must look for tomorrow and the future. We natives and whites let us put our hearts and minds together, and may the great spirit guide us to whatever way we can find to save Mother Earth. It gives and it provides so let us keep it clean as it was for generations, and work together without hurting one another. Amen.

Mary Ann Kkailther (Wollaston):

The Lac La Hache Band asked these people to come here to support us in opposing the uranium mining at Wollaston Lake. We asked the people to come here to find out more about uranium mines and how the people of Wollaston think and feel about the mines. We were against uranium mining since 1972. When mining companies and governments came to our community to talk about the mines they only told us good things about the mines. Things like giving our people jobs and using uranium for fuel and medicines. They never talked about uranium wastes that are dangerous

172

Many elders spoke from the heart about the need to protect the land for future generations.

to all living things. They never talked about how uranium is used in making bombs, but we know how dangerous it is. The Band was against Collin's Bay B-zone in 1981. We told the company and the government we don't want the Collin's Bay B-zone, but they did not listen. We cannot sit back and let them go on destroying our land and water. We live off the land, on animals, fish and berries. Some people and some newspapers say that people here are not against the mines. I know everybody here is opposing the mines. We had a lot of meetings where people spoke against the mines. The majority of the people signed petitions to the government of Saskatchewan to stop all uranium mining on our land. We are also waiting for Premier Devine to have a meeting with our Chief. We do not want Collin's Bay B-zone and surrounding deposits to open. This is why we are having the blockade on the road to the mines. We have to do this to make the government listen to us.

Wollaston Lake Support Group, Statement of Solidarity:

We thank you for the invitation to Wollaston Lake and your generous hospitality. We appreciate the accommodations and dance. We have come from all across Canada bringing our skills and love for the Earth. We are united with you in closing down the uranium mine and agree that it is time for direct action. We respect the need for local decision making all over the world and we ask for your guidance.

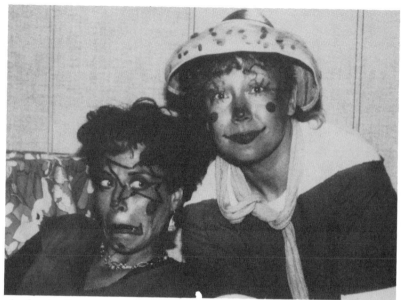

A short play portraying Eldorado's greed provided some comic relief during the serious discussions.

Elder Leo Medal (Black Lake):

We belong to this Earth, the land, because it provides and gives us what we need.

Work together as a whole group that is strong to help close this uranium mine. Make a group of people so strong that even if a rock hit them it would bounce off.

The rivers that go out of this lake go to every community around this area, so everyone should be concerned with what's happening to the land. Is it on account of money? Money is nothing for us. Sure we need money but not that way. We need money but the most important thing we need is the Earth. You young people must stand up and fight for the land. Sure the government gives us money but if we don't stand up and fight now we'll have no land left.

All the Dene nations in northern Saskatchewan, Manitoba and the Northwest Territories should come together to fight the uranium mines. Come together again at some place. Make every effort to fight these mines because it's our land. A long time ago we didn't know what the mine was. We lived on meat, fish, caribou, moose, rabbits, everything like that. How are we going to do that if the land and water is spoiled? What can we get to drink when

174

our kids are thirsty?

I wish the mine was not open. I care for my grandchildren and my grandchildren will get kids. Those young people think they will always get rations and family allowance.

It's true. It will not be like a long time ago again. We used to use a tent and maybe a log house. We didn't know what houses were. We used to live away from the town. We used to stay in the bush. Now nobody is in the bush. They're gonna take everything away from us. Nobody will know how to work like long ago. When I was a kid I used to wear caribou hide clothes. I thought it was nice. Now young people laugh at these clothes. Now they have skidoo boots and suits. We used to use dogs but nobody knows how now. Now young people laugh. Skidoos are not safe and cost money.

And the prospecting too, it did quite a bit of damage to our trapline. Also, about 20 years back we never heard about cancer. But now people are dying of cancer. It's probably because of the food they're getting from the south too. Before when we lived off the land we didn't have any store bought food. People were not suffering that much at all. There's a 30 year old woman in this community suffering from cancer, only 30 years old. What has caused that? And on my trapline I have seen a lot of dead fish in shallow water. We never saw that before. It could be because of those resource officers planting poison to get rid of wolves. They say there are too many wolves.

There was a guy named George Mercredi[4] involved in the uranium mine and he had a good job. He says he was raised around this area. He wasn't any better than any of us. Now he looks at it this way. He got his job and made a lot of money. Where he lived and where he was raised up, it's only his camp that he left behind. With all the money he made he's been living in the city. Us people, we have to realize that we eat what has been given to us by the great spirit and this land is all we need. I've worked in white man's camps before and I say to the young people that are around here right now, you'll never fit in the white man's world and we don't belong there. We belong to this Earth, the land because it provides and it gives and it shows. I never held a pen or spoke English. I managed through good and bad times and I'm not any better than anybody in this room right now.

Elder Louis Chicken, Senator FSIN (Black Lake):

I want to support the people that have talked and also you white people that came here to help us out, and I'd like to express my mind too.

175

The Band Hall was filled to capacity during the long meetings.

I've been around for a long time now. I remember way back in the 1920's when if a person had matches, bullets and tea that was a lot. A person could make a living out of that. Those were the main items.

When I was a young man people used to be happy the way they lived. People used to love one another. Everywhere the traditional way was going on. Hand drum, hand game. Everywhere people were happy the traditional Dene way. There was no fighting until 1940 when people learned how to make home brew. The old way of living was changed because liquor steals your mind and your spirit. People drink and fight over liquor. You don't know what you're doing. There's suicide and death over liquor.

We used to be hunting and moving all the time to feed our family. Now there's not many caribou around. They're not that close anymore because white people are chasing them away from us. Now our children are going to school to learn. I think children instead of learning white should be learning about our culture. They think they're smart but they're not. They should be learning

176

Translations were made from Chipewyan to English and vice versa.

to live off the land, listening to the elders.

All the people at this gathering are as one mind. We must work together. I want to thank all the people for coming to tell what wrong is being done to the land and culture. Every community has a right to speak up about what is right for them. The government should listen to us too instead of neglecting us all along. We have a right to defend our land. And we have the right to speak up for our own people. For the good of our people. If the government was listening they could understand what we are saying. All leaders should be listening to the people.

The first time the white men came they were not that rich. Now the government is very rich because they make money off our land. All Indian Bands should be here supporting the Wollaston people. The people in the north have been struggling all their lives. Now wastes are slowly killing animals and fish. We the people want everybody to live happily. When we see other people having problems we want to help. Why can't they do the same thing? In 1945 a lot of people died because of something from uranium. We

Each evening there was a dance.

don't want so many people to die from bombs again. I'm asking for all the young people of Wollaston to support the blockade to stop the mine.

You people who are writing down what the people are saying here about the mine, I hope you put it on paper so other people and the government will hear why we're having this gathering here.

I want to thank all you white people who have come this far to help us. When God made this world he made everything different, but still white people and Indians are human beings and we should help each other.

The mining is causing a lot of damage. I don't understand why the government is doing that to destroy our people and our land. It's not right to do that. We have a right to fight for our land and our culture. We should all fight together and stop this uranium mining. If we speak up maybe the government will change their minds.

Elder Tony Dzeylion (Wollaston):

I know this uranium mining will affect us here. I've been going to meetings down south. We are fighting so this mining will stop. There are a lot of people here from the outside and I know they're trying to support us, especially these white people that have come

178

Two of the "Northern Eagles," a Wollaston rock music group.

from far away. Just to see those people travel so far to come here to help us makes us really understand that this uranium is a bad thing that is happening here. A lot of us understand what uranium really means now. We'll try as hard as we can to stop it even though it might never work; us older people probably won't see the damage to the land but our children will, so we're not going to stand back and see it destroyed.

Pierre Alphonse (Black Lake):

There's 10 old people, myself and Billy Bouvier who come to Wollaston from Black Lake with empty pockets. But we've got feelings for these young Dene people now growing and not know-ing what's happening. We have to get our heads together like a great bird that flies high in the sky and sees us. Our lake that we live off is being poisoned and it will poison us because we live off the lake and because the poison will kill our food. They are killing off what we need to make a living.

If we know our traditional life we can speak against those that hurt us. But because we don't understand what the Earth means to us, we don't know where to start. We are part of the Earth and the Great Spirit gave us four legged animals to live on. I want the elders to come together and teach us the traditional ways so we

179

know what the Earth means to us.

Ron De La Hay (Fort Qu'Appelle):

I'm from Fort Qu'Appelle in southern Saskatchewan. For the last 2 or 3 years myself and a number of other people in Fort Qu'appelle have been working to try and stop uranium mining and the damage it's doing to the environment. We've been trying to stop the building of nuclear bombs from uranium, which could destroy the whole world. When we heard about your protest against the development of the mine up here, we were very interested in helping you, and managed to raise some money to send up from various members of our group and from church groups that also sent up some money. I hope it helps you to know that there's other people here in Saskatchewan that want to help you.

Michael Fitzsimmons (Pinehouse):

I'm from Pinehouse and we just spent 3 days in Prince Albert trying to raise support for the Wollaston Lake people. We got some people to come up here for this gathering and we also raised some money from the people who wanted to come but couldn't. So we want to give this money that we raised in Prince Albert, it's about $220.

Joanne Young (Zurich, Ontario):

"Eldorado" means "City of Gold" in Spanish. When the Spanish first came to North America and South America they landed in Mexico, and the native people in Mexico had a great deal of silver and gold. The Spanish people took the silver and the gold from the native people in Mexico who were called Aztecs and Mayas. They destroyed their civilization, they destroyed their cities, they murdered almost all their people. There are very few left today. This is what the "City of Gold" did for the native people in Mexico.

The name Eldorado then, should be a warning to us. The second Eldorado is out to get as much money from this land as they can possibly get. They don't care a hoot what happens to the people who are living there.

My own experience with Eldorado is very personal. My husband worked at the only refinery for uranium that we have in Canada. After they dig the uranium out of the mine on your property, they send it to Port Hope in Ontario where they take the uranium out of the rock. In Port Hope the land has been spoiled because the rock that is waste is left lying in dumps around Port Hope and it turns all the land bad. Then they use the uranium to make bombs.

My husband, when he died in 1956, had never been sick. He

180

Most of the adults in the community attended the meetings.

was a very healthy man. He was 34 years old. He had been working at Eldorado for 4 years. Shortly after our fourth baby was born he got quite sick. He complained of pain in his back and the doctor said he strained his back. He seemed to be losing a lot of weight. He felt very sad, he didn't want to do anything anymore, he just sat around. They put him in the hospital and did tests, but they couldn't find anything wrong. After he'd been unwell for about three months, he began to get lumps on his skin. Six months after he first began to complain about the pain in his back, my husband was dead.

When he died he was covered with lumps and the doctors said that he had lumps all through his insides. He inhaled, breathed in a lot of radioactive dust. He told me about that and I know it would have happened because he never would have told me if he hadn't known it happened. But the officials at Eldorado have always denied it. They say that the death of a person like my husband at 34 years of age is perfectly normal.

When Bill died, I thought that was just one accident that happened and that it would probably never happen again to anyone. It's only in the last 10 years that I realized thousands of people have died the same way my husband died. And it is the policy of our government to deliberately lie about the effects of radioactive dust and radiation. They planned it for a long time, they pass laws that protect them. I've tried everything I could to change their minds, to get them to play fair. The only thing I've been able to do that works is to protest non-violently by disobeying the laws that protect these dishonest activities of our government.

181

The Wollaston Lake Support Group singing "It Isn't Nice."

It Isn't Nice

It isn't nice, it isn't nice,
it isn't nice, it isn't nice.
It isn't nice to break the law.
It isn't nice to go to jail.
There are nicer ways to do it
but the nice ways always fail.
If that's the price of freedom,
I don't mind, I don't mind.

It isn't nice, it isn't nice,
it isn't nice, it isn't nice.
It isn't nice to block the road,
It isn't nice to go to jail.
There are nicer ways to do it
but the nice ways always fail.
If that's the price of freedom,
I don't mind, I don't mind.

When I was the age of most of the people in this room I never thought I would ever be proud of having been arrested and been to jail for protesting against the nuclear industry. But today I'm very proud to be able to stand here and say I have.

Maisie Shiell (Regina):

The Saskatchewan government, back in 1980, accepted a proposal from Eldorado to build a uranium refinery just outside Saskatoon at a place called Warman. This is a little bit of a success story for us because the people organized so strongly that the proposal was finally called off.

There was a government Inquiry about it. I've been involved with a number of Inquiries and always the mines have gone on. But this time the people were Mennonite and they were farmers and farming means a great deal to them. They came from Russia in the early part of the century because they wanted to get away from war. They were furious when they heard that the uranium refinery was going to be built in their district and that uranium is used in weapons. These Mennonite people organized. They went around to meetings all over Saskatchewan and asked for support. This was how so many people turned up at the Inquiry. Of course another reason was that it was so close to Saskatoon where a lot of people live, but it was the determination of these farmers who are very, very, very much against war, they're pacifists.

It's good for the people of Saskatchewan that the uranium refinery wasn't built, but this government is talking about it again and they are building another one in Ontario instead of Saskatchewan. So there are people and places that are still suffering from this. We can stop it in one place but then they'll move to another, or to another time.

Chuck McCallum (Ile à la Crosse):

I came up here to Wollaston to support the people in their fight against the uranium mine that is opening up here because I believe that there should be alternative ways of creating employment for the people. These mines create employment and revenue for the country but it's a very great danger. Spills have happened in Key Lake and other mines where they've said these mines are safe and nothing would ever happen. Nothing is infallible in this world. If it's manmade there's always something that could go wrong. The Cigar Lake uranium mine is supposed to be 10 times as rich as Key Lake which was supposed to be the richest in the world. The radiation from the uranium will go into the air. There will be seepage into the rivers and lakes and it's going to poison the fish.

183

A lot of birds and animals feed off the fish and so do we, the people of the north, and so do many people in the south.

We've been fighting for our rights for a long time. The treaties were a scam. Chiefs in the older days didn't know the system. The system is not here for their benefit but for the multinational corporations. The north is the last traditional land we have. Everything else has gone to the system. A lot of our people are caught within that system. They know very little of the dominant society system and very little of the native system, so there is a state of confusion. I can't blame my people for not being aware of the dangers of this uranium.

Mary Rose Robbilard (Black Lake):

I was a Counselor before and talked at meetings throughout Saskatchewan. Of all the meetings I've attended this is one of the best. You don't hear people arguing or talking back to one another. All of the Bands from here, Black Lake, Lac Brochet, it would be good if we went, the leaders, the Chiefs, to the blockade tomorrow. It would be good to round up all the Chiefs to go across to help us. Because the white people got a lot of ways they can turn things around. So if we got our leaders in there, we'll be much stronger for the blockade.

We Are People

What does the life of people mean to you?
Do you care for people and animals?
Do you think people and animals are not
important around Wollaston area?
People's important food are animals.
Animals' important food are animals.
Animals' important food are on the Earth,
under the Earth and in the water.
We are as human as you are.
We eat, we drink, we talk and we love.
Don't let us suffer for nothing.
But if you are opening that mine,
we will not stand around
and let you open it
without putting up a fight.

– Elizabeth Joseyounen (Wollaston)

The old medical clinic in the center housed many of the outside supporters. The houses at the right are known as "teachers' row."

Elder Abel Denedchezhe (Wollaston):

Us people we have our own minds to express what we think about the future and our own children. We don't just close our eyes and keep them shut. I'm asking everyone, especially the visitors from the outside, to help us speak up and fight as long as you can.

Benjy Denechezhe (Lac Brochet):

We came here to help one another. We have lived in this land for generations. We cannot let uranium mining go on any longer! It destroys our water and affects fish and wildlife. We can simply not let mining companies destroy our land just because of money. Money is not the only thing in life. We have proof that native people have lived off the land. Some people here have grey hair but they never held a pen or spoke English or had any education, but they survived and raised their families off the land. We native people all came to give each other a hand as a group and I am happy to be here.

Before it's too late we should do something about the mining across this lake. Sure, they'll give us everything we need, phones too, anything. What happens if the land is destroyed? Where can

185

we go? Now is the time to fight. Let's all stand on our own two feet and fight back. In the end there will be nothing left for us if we don't fight back now.

Back home our culture is a traditional way of life but slowly and surely the traditional way of living among the people is fading away, it's not strong. We have to know that way of life, how the native people have survived. From my home town there's some kids that are about 10 or 11 years old that don't know our language. It's slowly fading away. We have to teach the young ones how we survive. If a government comes up to us and says, "Where is your traditional way of living?" we have no proof. We have to have the language with us all the time so the treaty and native rights will be strong.

Sarazine Josie (Wollaston):
What I would like to say is about the mines, I know it's not good. It destroys our land, our animals, what we eat. Us young people should think about our future, if we have kids. And we should go across to the camp and help stop that uranium mine.

Marie Adele Josie (Wollaston):
When uranium mining first came George Mercredi was translating. He told us we would have a good and wonderful life in the community from uranium mining. He did not tell the people about the dangers it will cause.

Because of lack of security at the mine site many of the mine workers are introduced to drugs such as marijuana. The people that work there bring bad drugs. No matter which way you look at it the mine causes problems for the people.

Brian Ratt (Ile a la Crosse):
I'd like to say that in this community people are sticking together better than any other place I've been. The government always comes and gives a little bit of money to someone over in the corner but then everybody else doesn't have anything. The white men do that because they're trying to divide people. Jealousies and everything start up. Or they bring in a case of whiskey for the people and say, Don't listen to those other people over there they're crazy. I've seen it and I've heard it.

It's just now that I've seen so much solidarity among the people. Last night a lot of people had a good time. Where I come from people have to drink first to have a good time like that. But last night I saw old people and young people dancing and having a good time without booze.

186

I didn't know what to expect from Wollaston because in Pinehouse they let the uranium mine in. A lot of times I thought the people there would even fight harder because the road is right beside them. But a few people get paid off and jealousy starts up, then people get confused and the government just sits back and laughs at the people fighting amongst each other.

We've got to learn that when a white man comes along we've got to watch him because he's a cunning man. When he came to this country he was starving. He didn't know where the hell he was. He even thought the world was square. But we always knew the world was round like the moon and sun. But we were so kind-hearted and the white man took advantage of us.

There's a little story about when the white man first came here. He was walking along a road and saw an Indian sitting on a log and he asked the Indian if he could sit down. The Indian said, "Of course, sit down." But pretty soon there were a lot of white people sitting on the log and the Indian kept moving over. Pretty soon the Indian was sitting on the ground.

We have to pray to the Great Spirit and ask for all the strength that we need because we're a very kind-hearted people.

Elder Bart Dzeylion (Wollaston):

For the people that are going across to the blockade, our religion is what we need. Our religion is different compared to some of us. It's good to have people that know the bible. Everywhere we go us people we have religion, belief in God. Everywhere we go we need that strength to keep us.

Billy Bouvier (Black Lake):

I was born and raised in Black Lake and I've been living off the land for most of my life. I've been eating fish and animals here that live off the water and the water gives us life. I came here because I think of the future. I'm a young man and I've been working in northern Saskatchewan on the people's alcohol and drug problem. It's a problem that the white man gave us. They destroy our land and they destroy our people. We have to work as a team and help each other for the future generations.

Adele Ratt (La Ronge):

I want to say something and I think I'm speaking on behalf of the rest of us that are here to support you. We'll do the very best we can to help you stop uranium mining up here. A lot of us here have travelled far and have done a lot of work already in our own communities, spreading the word around all over, in Canada, the

187

United States and overseas.

Many of us have worked against uranium mining and in the Native rights struggle for many years. We've been persecuted for speaking the truth. Sometimes it has been really hard and frustrating. I just want to say, and I think the rest of the people here from the outside share our feelings, that the strength and power that you people have shown us in these meetings has given us a lot of strength. We always have to remember the Earth, that we are only a small part of the Earth and our power comes from the Earth. Whenever the Earth is hurting, the people are hurting.

We also have to remember that there's no room for hate in our lives. The enemy, this thing that we are fighting against, does enough hating for the rest of us. So we must always remember that we do these things out of love for the people, and I think I can say for the rest of the people here that our hearts and our minds are one with you in this struggle.

How I Wish

How I wish,
I could keep the image of the beauty
of Wollaston Lake,
pure land, air and lakes.

How long can I keep Wollaston Lake pure
before uranium mining destroys the purity
of land, animals and lakes?

What a shame!
How could anybody in their right mind
destroy a beauty such as Wollaston Lake?

– Sophie Denedchezhe (Wollaston)

Sophie Denedchezhe and her son.

At Wollaston:
Does The CBC Have No Common Sense?

Young Adam stayed behind to watch TV
and avoid the long meeting
and I stayed behind because I had an infected toe.
Good reasons.

"Is that a fish spear," I ask, "How does it work?"
He thrust the spear at a wooden rail
and steel trap jaws snapped over the imaginary pickerel.

We stood while we watched TV –
CBC Northern Service from Newfoundland.
Has the CBC no common sense?
Weather forecast: rain, fog and drizzle,
rain, fog and drizzle, rain, fog and drizzle.
"Look at the good side," said the announcer,
"Ottawa just put a tax on suntan lotion."
Adam glances at me, "What is suntan lotion?"

Adam is about 14 I think, just my daughter's age.
Adam is dark as shoe leather.
His muscular arms brandish his spear.
Dark hair held by a band. Black leather vest, Levis.
He lights a cigarette. I am grateful he does not offer me one.

"Oh, I've seen this picture!" he shouts,
"That guy gets shot in the neck, you'll see."
The enemy circles around the miners.
On TV the miners are on our side, the good guys.
The shooting intensifies.
Yes, he gets shot in the neck.
"Here comes the cavalry!" Adam shouts.
The miners are safe. The enemy is gone,
cowards all against a determined minority.
This time the enemy is the Mexicans.
I breathe another sigh of relief. They could have been Indians.
The power goes off. I am grateful.
Does the CBC have no common sense?

– Jack Ross (Cooper Creek, B.C.). The poem expresses his experience in the Wollaston medical clinic while the June 11th meetings were taking place.

Waiting for the boat to the campground.

June 13, 1985:

The community meetings in Wollaston concluded. About 150 people moved from the community across the lake to camp at the Umperville River campground. An evening feast was held by supporters in honor of the Dene, followed by a traditional round dance outside the elders' tepee. The first ever pre-dawn, multi-cultural, non-violence workshop was held, complete with Dene/English translation.

191

At the Umperville River campground.

192

Photo: Marie Tutt

People living far from each other had the opportunity to share experiences.

194

Everyone gathered early in the morning in the drizzling rain for the final meeting before the blockade.

The Blockade, June 14–17, 1985

June 14, 1985:

Chief Hector Kkailther at the Umperville River Campground, just before going to the mine gate, June 14, 1985:

I tried my best to get the government and the mining companies to discuss how it will affect the people for years to come. I tried my best to have a meeting with the people involved in uranium mining, and the people looking after it, running it, the ministers. It seems like those people are only looking out for the money. That's their main concern. You people from the south came a long way to support this gathering. We can all work as a group and become one strong group, depending on one another. You know how the governments are, it's just like one day for them. We have to keep on bothering them all the time, coming back, what we started, we have to keep on doing all the time, no matter if we're alone or whatever. We have to keep on bugging them until they close the mine. That's what we're after.

195

Chief Hector Kkailther speaks to the group.

These governments and the people who are owning the mine in the first place they should have settled with the people and discussed things over and talked about how it should be run but they never did that. They went on their own without letting the people know. But they make money out of our land. They make thousands and thousands of dollars out of our land. They damage the lake, the land, everything, animals, fish. We're left behind with nothing. All the money they make, they go back some place with it and we're left here behind with nothing. That's why we're gathered here today on that concern. It's not too late if we fight back. The land is still beautiful. Look around. So let's keep it that way. Let's fight them with our own two feet. With strong hearts and minds we can work together. They're human beings, they eat same as us, they got brains same as us. They breath the same air. So no matter what, let us all work together and stop that uranium mine. No matter what let's all work together, make a strong group. If we keep on bugging them hopefully they will close the mine so the land can still be beautiful for years to come.

196

On the way to the mine gate from the Umperville River campground.

Arriving at the mine gate.

At 11 in the morning a caravan of vehicles moved to the mine gate. When the cars, trucks and vans holding about 150 people arrived the gate was closed. Eldorado security vehicles were parked a few hundred yards inside. After a short time a group of four Dene elders and an interpreter walked around the fence and up to the Eldorado staff. The elders requested a meeting with the mine management. The request was refused.

Chief Kkailther then phoned Saskatoon to request a meeting with mine manager Mike Babcock, who finally agreed but would not set a date or place.

197

Elders on their way to speak to Eldorado security.

Blockaders speaking with CBC La Ronge reporter Ray Fox (center, wearing headband).

198

Wollaston Elder Bart Dzeylion being interviewed by CBC Saskatoon reporter Dorothea Funk.

Watching and waiting.

Chief Hector Kkailther, Mine Gate, June 14, 1985:

I'm gonna phone that Babcock, he's the manager in Saskatoon and try to get him today to come to this blockade. I want a straight answer to what he thinks and also we want to tell him what we think. So I told these guys I'm gonna phone him after lunch and so I'll try to get him here today somehow. I want him to come, that's the first step.

199

The Rabbit Lake pit and mill buildings in the background.

The arrow marks the point where the blockade took place.

All eyes were on the RCMP when they went to speak to mine

The two local Royal Canadian Mounted Police (RCMP) Officers, Ross Reynolds and Doug Hardy, arrived at the blockade and were invited to speak to the people. They refused, claiming they were impartial. However, Officer Hardy took many photos of the blockaders while conspicuously ignoring Eldorado security.

Chief Hector Kkailther (HK), Stephanie Sydiaha, and RCMP Officer Reynolds, at the mine gate, June 14, 1985:

RCMP: If you want I can act as an intermediary and pass the message on 'cause I don't know if the mine itself is in contact with Mr. Babcock at this time but I'm sure he will be calling.

HK: OK, just tell him to phone me right away.

RCMP: I'll pass on the message.

HK: I tried to get him to talk to these people.

RCMP: We're not disputing your right to be here. It's just that we would like you to remain on this side of the fence.

HK: This side of the fence, that's one thing too.

RCMP: Present legislation that we are bound to enforce says that is lawfully leased land and it is under control of Eldorado personnel.

202

officials . . ., . . . and then left.

Now until such time as that dispute is settled, I think it would be better off for everybody if they just remained calm and remained on their respective grounds where they know they're not going to get into any problems.

HK: But the one question I got is they should have told those people here. What about them coming here in the first place? We got our rights.

RCMP: There's no one disputing that you may have claims to the land, Hector, but as the present treaty stands, they are lawfully mining this property and under the authorization of the federal government.

HK: (Heavy sigh.)

RCMP: And until such time as that's changed there's nothing going to happen here.

HK: They should explain to the Band in the first step, before.

RCMP: There's a lot of things that should have been done.

HK: That's what I told them.

RCMP: But the thing is, the fact is that at the present time they are there legally and until such time as the legislation or the legislators see fit to discontinue this type of industry, all industries, then we can all go and live in the bush because nobody's going to have any money.

203

Riot shotgun in the RCMP vehicle, placed between the driver and passenger seat.

Stephanie Sydiaha: Maybe this is an attempt to change the legislation.

RCMP: Through lawful means I have no objection.

Stephanie Sydiaha: They've been trying to change the legislation through lawful means. Right?

HK: Right!

RCMP: I haven't seen any real problems in other provinces with settlement of land claims and involvement of the native people. Now if there's a particular problem in this province I don't know. But land claims have been settled legally in other parts of Canada and I fervently hope it can be done here so that we can stop all this. Because I have, personally, other things that I have to do. But the thing is nothing's going to change until the law changes. And we are here to enforce the law as it presently stands. And we hold nothing personal against any of you people, or the mine.

HK: Or the mine, yeah, well, we'll see what happens.

RCMP: All right I'll pass the message on to the supervisor.

HK: All we want is a meeting with Babcock. He got no problem to come up here. I know he can fly his jet up here. To our Indian people it looks like he doesn't want to look at us. All we want is our land back.

RCMP: I can't speak for Mr. Babcock just the same as I don't speak for you.

204

In front of the mine gate.

Wollaston Elder Abel Denedchezhe cooking a fish over a campfire on the road.

The road was turned into a place for celebration and living. It was a heart-warming sight. A cooking fire was placed in the middle of the road where caribou meat and lake trout were prepared. Large stones were pushed in front of the gate and used as comfortable stools. People sat calmly as if there wasn't a worry.

The vast majority of blockaders were Chipewyan residents of Wollaston Lake. Supporters came from the northern Dene and Cree communities of Lac Brochet, Black Lake, Pinehouse, Southend, La Ronge and Ile à la Crosse; and from southern Canadian communities in BC, Alberta, Ontario and Quebec.

In the evening elders stood around the fire and sang ancient songs while beating a caribou-skin hand drum. At the same time young and old people participated in a traditional round dance – with the elders in the centre. On the side of the road about 25 tents were pitched.

In sharp contrast to this peaceful scene was the ever-present dull roar of mine machinery and the vigilant Eldorado mine security personnel. But the sound of the drum and the singing carried well into the Eldorado camp. Workers who wanted to go to the gate were told they would lose their jobs if they did. Some

206

Sarazine Josie and tents on the side of the road near the mine gate.

of the mining company staff appeared to be anticipating a barrage of flaming arrows over the gate at any moment.

Meanwhile a support vigil took place at the Eldorado Nuclear building in Toronto. It started at 8 in the morning and lasted 18 hours. The protest included street theatre and a slide-show using the side of the building as a screen. A small protest also took place at the Canadian Embassy in Dublin, Ireland.

From right to left: Adele Ratt, Jake Badger, and Stephanie Sydiaha.

Protesting outside the Canadian Embassy June 14, 1985 in Dublin, Ireland. Right to left: Rosemarie Rowley and Tom Kenny.

Letter From An Indian In Prison

I cannot say "I".
The person who is writing this
Is not me, but
An adjunct of the State,
My private prison officer.

I am sending this out to you
From the tyranny
Of petty officialdom
In this tired officialese
Hoping you-who-is-in-me
Will see the spirit
Cowering behind the blowtorch
Like an extinguished candle
On which an ember glows
You are the audience
Who will put that ember
In your mouth
To cleanse yourself, and I
Am one who can speak to you
Out of suffering.

My vision has robbed my mouth
My taste is of stale dry bread
But I do not hunger
For the white man's feast
But call it famine
I do not long
To share your tap
Of spring water and call it truth
Only to find
A metallic taste
Where you have poisoned our inheritance
Our lands are robbed
Our children full of disease
Our animals wasted

A cup that cheers
Would need a brighter day
A day that calls us together
Is what I long for
In writing to you.

– Rosemarie Rowley, Dublin, Ireland, 1985

HONOUR MOTHER EARTH

Stop the Cycle Where it's Sound...

Leave Uranium in the Ground!

VIGIL AND STREET THEATRE

In Solidarity with the

Wollaston Lake Anti-Uranium Rolling Blockade

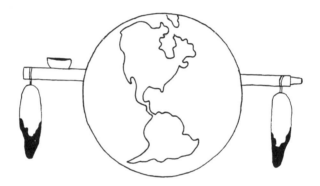

FRIDAY, JUNE 14, 1985

Outside offices of Eldorado Nuclear Ltd.

100 University Avenue, at King

Vigil: 8:00 a.m. to midnight

Street Theatre and Speakers: 8:00 a.m., 12 noon,
and 6:00 p.m.   Slide show: 9:30 p.m.

For more information, call 964-0169 or 537-0438

SPONSORS:
Native Education Society of Toronto
Canadian Alliance in Solidarity with Native Peoples
Alliance for Non-Violent Action
Campaign for a Nuclear-Free Ontario
Toronto Nuclear Awareness

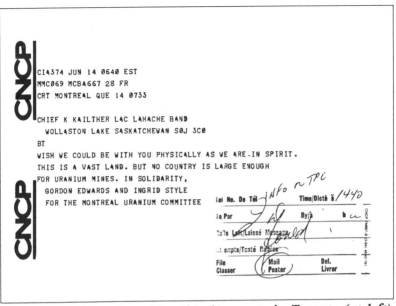

CIA374 JUN 14 0640 EST
MMC069 MCBA667 28 FR
CRT MONTREAL QUE 14 0733

CHIEF K KAILTHER LAC LAHACHE BAND
    WOLLASTON LAKE SASKATCHEWAN S0J 3C0
BT
WISH WE COULD BE WITH YOU PHYSICALLY AS WE ARE.IN SPIRIT.
THIS IS A VAST LAND. BUT NO COUNTRY IS LARGE ENOUGH
FOR URANIUM MINES. IN SOLIDARITY,
    GORDON EDWARDS AND INGRID STYLE
    FOR THE MONTREAL URANIUM COMMITTEE

*Poster advertising June 14 solidarity events in Toronto (at left).
Telegram of support (above).*

211

RCMP Officer Doug Hardy confiscating a can of spray paint.

Photo: Sid Paschen

June 15, 1985:

The RCMP, armed with handguns and a camera, wandered freely back and forth from their vehicle through the blockade camp to the Eldorado security. Officer Hardy, with Officer Reynolds backing him up at a comfortable distance, confiscated a

Smoking caribou meat at the blockade camp.

can of spray paint near the mine gate. In a few moments the police were surrounded by blockaders and a tense exchange of words took place before the RCMP entered the mine area.

At the meeting which followed, the elders resolved that the RCMP be requested to leave their weapons and camera in the police vehicle and be escorted by an elder when passing through

A line of blockaders.

the camp. The officers soon returned, but when an elder politely requested Officer Reynolds' cooperation, he looked nervously about at the crowd, shrugged his shoulders and said, "We'll come and go as we please."

The officers returned to their vehicle, followed by almost everyone in camp. A line of people stood about 15 metres away waiting for them to drive off. Police Officer Hardy took the opportunity to take some group photos. A group of blockaders responded by singing the American Indian Movement (AIM) honor song, using the hood of a nearby truck for a drum. A silent line of people stood with fists raised skyward while Officer Hardy's camera clicked away.

After the RCMP left, the blockade was physically bolstered by the creation of a southern 'security gate' about 100 metres south of the Eldorado gate. It was made by putting boulders and trees across the road, leaving only a small space for slow moving vehicles to pass. The gate served to keep guns, alcohol and drugs out of the blockade area.

There was much activity in camp with people still arriving. News arrived that trucks bound for the mine were being held back at La Ronge by the RCMP.

The RCMP photographing the line of blockaders. Below, a block-ader takes a picture of the RCMP.

Meeting of an affinity group, sometimes called the "infinity group."

Carrying a rock to put in front of the mine gate.

The road in front of the mine gate.

The blockade outhouse.

217

Mary Ann Kkailther in the centre.

Mary Ann Kkailther (MK) interviewed at the mine gate June 15, 1985 by Dorothea Funk, CBC Regina:

CBC: I'm speaking with Mary Ann Kkailther and she's with the Wollaston Lake Indian Band and the wife of the Chief. You've had this protest going now for a day and a half. How do you feel after this length of time and about the camp that you've set up?
MK: So far this thing is going good and a lot of people from Wollaston are coming and almost everybody's here from Wollaston.
CBC: Now what's happened is that during the time you've been here it appears that no trucks have gone through. What do you think that's meant for the mine?
MK: I don't know. Maybe they're just trying to see what happens if the trucks don't come through. They probably think we might leave or something like that.
CBC: Will you?
MK: No! We'll be here waiting for them.
CBC: Have you any idea when the trucks might come through?
MK: No.

218

Singing and using the hood of the car as a drum.

CBC: But you're going to stay here, you've set up camp and you're staying here, is that right?

MK: Yes.

CBC: What would end the blockade for you? On what basis or under what conditions would you end the blockade?

MK: I think if the Premier and those people at the mine, the bosses should meet with the Chief and they should discuss how they should stop uranium mining.

CBC: That's what you want, a stop to uranium mining?

MK: Yes.

CBC: How many people do you think are here from Wollaston?

MK: It's hard to say. We went back yesterday and all we could see was just some kids at home. Most of the parents and the teenagers are here and if people are back home, they're mostly baby sitters.

CBC: I understand you've got tents here and you've even got an outhouse set up so it looks like you've really organized yourselves. One thing I've noticed is that you've set up a gate at the entrance to your camp. What's the reason for that?

MK: It's just to have safety for the people here and to make sure

219

Cooking fish on the road.

nobody comes into the camp with drugs or guns and things like that.

CBC: What do you mean for the safety of the people? Who might bother the people?

MK: It's hard to say. You don't know what those people are up to.

CBC: Who do you mean by "those people?"

MK: From the mine and maybe the RCMP.

CBC: Now you've been here for more than a day and a half. Do you think that's been long enough to maybe stop production at the mill? Or is it hard to say?

MK: It's hard to say.

CBC: How do you feel about the fact that the trucks have stayed away for a day and a half? Has that ever happened before to your knowledge?

MK: Not that I know of.

The King of England once told the Indian people if ever they were in distress to hang the flag upside down and he would come to their aid.

221

Looking south at the security gate placed about 100 metres south of the Eldorado gate. RCMP Officer Reynolds, in plainclothes on the far right, is speaking with CBC reporter Dorothea Funk who is only partly visible.

June 16, 1985:

The mood changed dramatically. The RCMP arrived in plain clothes with no guns from inside the mine site. They had flown across the lake from Wollaston and landed inside the mine area. A short private meeting was held with Chief Kkailther and a small group of elders.

Chief Kkailther announced that Babcock agreed to a meeting at 4 in the afternoon Monday, June 17th, place undecided. Soon afterwards the RCMP announced there would be no traffic in or out of the mine site until after the meeting. It looked as though the people's views would finally be heard. The stress of dealing with arrest was temporarily relieved. More and more people continued to arrive.

The following statements are from June 16th.

222

Chief Hector Kkailther giving news to a group of blockaders.

Chief Hector Kkailther:

At about 4 tomorrow I'm meeting Mr. Babcock, the Manager from Eldor. So it looks like we're going to meet him tomorrow sometime. Before that I'll try to get contact with the other Chiefs and also the southern Chief and explain why we're gathering here and that we want to know what they think too. For me personally, I don't want anyone to get hurt at this blockade. And it seems like there will be no trucks anyway until after we meet with the Manager tomorrow. Besides that we want the meeting right here at the gate here or inside at the mine. What I'll do is phone around this afternoon and try to get other Chiefs involved in what's gonna be. Then we'll start from there.

When I have meetings with people like the Ministers and all those people we have had meetings with I keep on telling them, "You white people, you came to our land, this is our land, we can't

223

The main cooking fire at the blockade.

always follow what you say. You have to come to our side and listen to us, not us listening to you, you have to listen to us." I keep on telling them through the meetings. For the meeting tomorrow we have to make sure everything is together because the white man can change things around. All the elders get together and discuss this over. They have a lot of ways to look at things.

Today most of the people understand what is going on. But when the first treaty came up, if all those people knew what was going to happen a lot of things would have been changed and a lot of things of the white people would have been refused, like taking the land away and pushing us away.

It would be good if all the elders get together and discuss this over. We don't know how long we're going to stay here. We can't play around with the food that we have because there won't be enough for everybody. We have to watch among ourselves. Watch everything around us. I hope things turn out good with the meetings. Whatever the elders say, they can't always have the white man's way. They have to come to us to discuss things over. Most people have families across. You have to take care on both sides, that's your responsibility. That's the kind of stuff we have to look out for.

We're all gathered here in a big concern. All the elders have to get together and discuss this over to be prepared for tomorrow. If

Photo: Marie Tutt

tomorrow doesn't work out, we can always come back. When the next time comes, we now what the white man did and we can learn from that.

We don't know what's gonna happen tomorrow. This is not the only time, we can't quit now that we're started. We have to come to something that's gonna be good for the future.

Elder Leo Medal (Black Lake):

Sure we have lots of wonderful things right now, but when I was a young man we had to make our own gunpowder. There wasn't much around.

Before the Chief never used to have to talk to any outsiders, just his people. But now today you have to talk to some people – big boss, governments. In the olden days it wasn't like that, you had to live off the land.

When the native people get hired they get the labor jobs. They get the dirtiest jobs they can find and they don't get much money.

We cannot give up now since we've come this far. This blockade, we cannot give up now. Some of us have jobs and business to do, but we can always come back. We can't give up right now and say it's hopeless. We have to keep on coming back. You never know, we go back and bring back a larger group.

The government, the guys with all the power, they are hard

225

headed, they can't take up the Indian ways. They got rich with our land, but where did they put the money for us. They only give us so little that we have to fight among ourselves. The government can't agree with the native people.

We Dene people we have our history and we always struggle to survive, but we have enough courage and strength and hope. To this day we still have the courage in each one of us to fight for what's right. I hope the creator helps us all the way to stopping this uranium mine.

All you young people, what you learn from the ways we're living now, don't give up. If you give up, we have nowhere to go, nowhere to turn, so you're the ones who have to keep on struggling.

Adele Ratt (La Ronge):

Alcohol is the first thing that they always come into our communities with and we have to understand that. They use it against us. They brought it in and introduced it to our people and they're still using it to this day. We have to understand that it's part of the war that is being waged against Indian people today all over the country.

We have to remember our history. We have to remember the past, and the way they've treated us, and that a treaty is a sacred

226

thing to the Indian people. When Indian people made treaties with the white people they understood that these treaties were contracts between one nation and another nation. They have broken that contract and now I think it's up to Indian people to stand up and say, "You broke that contract, now we want something done about it."

It's important for us to know that these things have happened in the past and that they still happen today, that they still speak with that forked tongue, that they lie to us, and they try to make deals with us when we sit down to talk with them. And they always try to make us come to them to talk. We have to make them come and listen to us. We've listened to them for 500 years and look at the mess we're in.

I'm young and I'm still learning a lot of things. I always try to listen to the old people and follow what they say but sometimes we young people are really eager because we're just learning and beginning to understand what is happening to us. Sometimes we want to stand up and fight. We want to do all these things for the people but sometimes we don't want to stop and listen to what the elders are saying. It's important to do that because the elders have lived longer than we have.

I've been to a lot of gatherings and I see a lot of good things happening. It's good that people come together like this and speak out. It's important that we do that all the time, not just once in a while. It's hard sometimes because I have children, but I see myself as a warrior too. In this day and age things are different. In the old days the young men were the warriors but these days all of us are being attacked so strongly that women have to be warriors too.

Sometimes there is a lot of confusion among the people. Sometimes the warriors are quick to jump up when they see a threat to the people, like yesterday when the police came in and took that spray paint. Right away the young warriors were there and that could have turned into a bad thing and hurt some people. It was OK the way it turned out but the police took pictures of people and they were watching us through their binoculars and they saw a person from Wollaston Lake spraying that sign over there that everybody didn't like. And now they could charge that person with mischief for doing that. It's good we have warriors, we need them but sometimes they need to listen to the elders about when to pull back and when to go forward.

I hope that tomorrow at this meeting something good comes out of it and that we won't have to stand up to them anymore, but for myself I don't see that happening. I don't trust them because they

The hand drum made of caribou skin has been used by the Dene people since time immemorial.

came in here and lied to us to begin with to get what they wanted, to get the land. I trust the elders.

Billy Bouvier (Black Lake):

I have to share my feelings with the people. I know why I came here for this gathering, because we have to save our land for the people in the future. I've been working with the National Native Alcohol and Drug Abuse Program throughout the northern Bands since 1983.

It's really important to be here with the people. I know there are people crying, crying out loud for help. I heard about this blockade in a letter from the Wollaston Band and I phoned for information and told them I want to come here to support the native people. I came down here and I plan to stay. But it's hard for me because I have to look after my kids and I have to look after the people in the north. I've been working for Wollaston for 2 years on the problem of alcohol and drugs. I know there's a problem because I went through it.

I believe in religion, spiritualism - the sweat and round dance. We have to continue and live with those things. And also God gave us minds to think of good things, not the bad things. He gave us eyes to look upon the land and look at the good things, not uranium mining. And God gave us a mouth to talk about good things. We're talking about the land here, not about the Rabbit Lake mine, we're fighting against it.

228

I've been travelling all over the place and across Canada where native people gather. I learn from them. Yesterday I went back to Black Lake and I have to pay my way back here to the meeting. The plane cost a lot of money but I don't care about that. The native people give me my life back.

I've been sober for 4 years and my mind is clear to think about good things not about bad things. I got into a lot of accidents before. I got alcohol out of my mouth - I don't want that in my mouth again. At this gathering we try to solve the problem. That's all I have to say.

Elder Louis Benonie (Wollaston):

There's been lots of meetings. They've always said stop the uranium mine, but they refused to listen, so we have to do this today. We can't always let the white man tell us what to do. All people from Wollaston want to stop the mine.

Jake Badger (Mistawasis):

The way I've learned in my teachings and understanding is that the land does not belong to any individual to claim. The land is a gift to us to take care of and protect. The land belongs to all people and when there is something threatening the land, like the way it is being raped up here, it should be a concern to all because it concerns all, not only the people from around here; because all over the world the uranium is affecting the land.

The yellowcake is trucked right through Saskatoon and sent to Port Hope, Ontario, down into the United States and overseas to make bombs. It's affecting the whole world. People died in Australia because of bomb testing. In the South Pacific, islands don't exist anymore, like what they call the Bikini Atoll. It was blown up and there is a big whole in the middle. A bathing suit was named after it.

A group of us just finished a walk across Canada. We stopped at Indian reserves and prisons, wherever we could have gatherings with the people. We listened to their concerns, the issues that were affecting them in their communities, the harm that was being done to them and the conditions that we are forced to live under. In Grassy Narrows, Ontario the people's whole water system has been poisoned by mercury. They can't eat the fish. There's even cases of children being deformed.

People are suffering from radiation sickness in places like Pine Ridge, South Dakota and Navajo reservations due to huge piles of radioactive mine tailings. The nuclear war process begins on Indian land. Presently 14,000 Hopi and Dine, as the Navajo are truly called, are being forcibly relocated from their ancestral homeland in the Big Mountain area of northeastern Arizona. Many say they will have to be killed before they will be removed. Some speak no English and never have been off the reservation in their lives. It is a barren, dry land. No white man wanted it until coal, oil, and uranium was discovered there. Multinational corporations and the U.S. government want the resources that lie in the sacred grounds of Big Mountain.

I understand that when they first built this mine they did not consult the people in this area, yet they say you have to wait to talk to their boss. How many more years do we have to wait to talk to their big bosses?

It's good that this is happening because every day those trucks aren't running it's costing them money and money is their God.

All across Canada there's people that are struggling. On the west coast, mountains are being stripped of their logs. In Alberta there's the oil fields. Here, there's the mining ventures. There's people that are struggling, standing up and resisting and the more we resist the stronger all the struggles are because it is the same struggle. Together I know we can be strong.

Rico (Toronto):

Where I just came from in Ontario five of the largest lakes in the world have been polluted and in northern Ontario many of the lakes are irreversibly dying from acid rain. The government is dragging their feet from election to election making promises that they would clean up the lakes. There have even been talks between the United States and Canada to get together and clean up the lakes. They've been doing that for the past ten years.

Pierre Alphonse (Black Lake):

We're not blocking the road to a uranium mine, we're blocking the destruction of the whole planet.

Rico holding Angus Kkailther.

230

Identity

The late meeting is over.
Exhausted, tense, anguished,
I am poised between rest,
nourishment and companionship.
I drift to the cookfire in search of food.
Rico is teaching my friends his Yoruba songs
– percussion, leader, chorus.
His deep bass soars. We follow.
I join the rhythms with a lid and spoon.
"Don't worry about a thing,
Everything is going to be alright."
I glance north at the guards' vehicles.
South to the police truck
and sing louder.
"Everything is going to be alright."
I finally say good night.
I have acquired a Yoruba accent.
Who am I?

My mind repeats the common sense message,
go to bed.
On my path the Chipewyan elders
are tuning their drums in the fire.
They begin the heavy syncopation.
They chant: the anguished wail of a people
who are dying charges the air
with a feeling more powerful than I can resist.
We dance. Always clockwise they tell me.
Did the elders always know about clocks?
Will their grandchildren know that
clocks had hands that turned?
I fall in behind Abel,
and try to mimic the studder step
that seems so easy to his many years.
Dancers come and go, but the drums continue.
Are the guards dancing? Do radiating molecules
dance as they penetrate the vital cells?
Emma Goldman said,
"If I can't dance I don't want to join your revolution."
As the sun rises I finally stagger to my damp, cold bed.
My mind echoes the wail of the chant,
and my chattering teeth keep time

to the memory of the beat
I still dance. I chant Chipewyan.
Who am I?

I give up on the cold blankets
and fall into a fitful sleep in the car.
My young Black Lake Indian friend Joe soon wakes me
with his impish grin.
Hey you awake yet? Can I sit in your car?
Can I hear Beethoven?
Who is Beethoven?
Who am I?

– Jack Ross (Cooper Creek, B.C.) June 21, 1985.

Gerry Paschen at Mickey's Camp (130 km north of La Ronge). His car was damaged from a collision with a deer.

A Blockader's Journal, June 12–17

By Gerry Paschen

Gerry and Sid Paschen from Edmonton, Alberta attended the gathering and blockade. Following are excerpts from Gerry's journal.[5]

When we arrived at the Umperville campground on June 12, 1985 we were surprised at not finding the large group we expected. While my son Sid and I unpacked our car and set up camp, Allen Quandt came over and joined our campfire. He told us that all of the volunteers for the survival gathering had been invited to the Wollaston Lake Indian Reserve.

Allan Quandt is a highly motivated individual and deeply

dedicated to the cause of stopping the arms race. He told us that he once skied all the way from his home in La Ronge to the Primrose Air Force Base on the Saskatchewan-Alberta border in protest against the cruise missile testing. I then told him my account of a demonstration in Edmonton against the cruise missile. We sat around the campfire half the night exchanging ideas.

While we were relaxing around the peaceful campsite the volunteers were socializing, dancing and discussing the "big event" at the Wollaston Reserve. Allan commented that if the press were to arrive at the campsite at that moment, they would conclude that the level of interest in the blockade was very low. He said that from past experience he had learned that the press capitalizes on non-interest to gain momentum against the demonstrators. He said the CBC were the only press who had arrived at the camp but warned that their facts regarding the blockade might come out somewhat twisted. He said, "Don't judge low attendance as a failure in a proposed project like the blockade but pay attention to the positive things that are going on in the community of Wollaston Lake." Never before had this Indian Band invited 30 people from outside to visit their reservation. Early the next morning, June 13, Allan left camp.

That afternoon an RCMP van pulled into the camp and several members of our party spoke with the two Officers. One Officer asked me if we were planning to take any weapons to the blockade. "No," I said, "only mosquitoes." He grinned at me and replied, "That is sufficient." He and his partner then left the campground.

Shortly after, boatloads of people started arriving from Wollaston Lake. A large kitchen was set up at the camp, and in the evening we had quite a feast of fish, potatoes, venison and vegetables. We then finished making a rope of four strands to hold the top of the tepee poles together. A fire was started and some drumming began, prompting several of the campers to stay up and dance all night. It was 2:30 in the morning and daylight was already breaking when Sid and I went to bed.

As it turned out, I stayed at the campground to keep an eye on things when everyone made the trip north to the mine gate. At one point during the day an American fisherman came by and gave us 35 fifty centimeter-long fish, weighing an average of four kilos each. I immediately gutted and cleaned the fish and then later put 10 on the fire to cook for those at the blockade. By early afternoon the first few people returned to the camp and reported that the blockade was working.

I went to the mine gate to deliver the cooked fish. What a

235

fantastic sight it was! A picnic table was in the centre of the road in front of the steel gate and there were several fires going, giving off clouds of smoke to keep the mosquitoes away. It was a very happy crowd gathered at the gate, eating, visiting, singing and playing. They were spread out east and west of the road and several of the children were playing frisbee. Later in the day tents were put up for those who would spend the night at the blockade.

I returned to the campground. Back there a new opportunity presented itself. I was invited by Mr. St. Pierre to travel in his boat across the lake to the Wollaston Lake Reserve. His boat was a flat-bottomed 8 meter skiff with twin outboard motors. I quickly grabbed my bedroll, a windbreaker, and a few other essentials, and boarded the boat.

A strong, cold wind was blowing as we made our way east across the lake. The lake was very choppy and there were still chunks of ice floating in it. I appreciated Mr. St. Pierre's skills as a captain. Finally we observed some houses on the horizon.

The sun was setting as we pulled up on the shore. We were blue in the face as we made our way up the rocky beach to Mr. St. Pierre's house. I could hear the sounds of his sled dogs barking and could see his children through the large living room window. The oil burner was on and I welcomed the warmth it gave off. Their home was simple but very clean and comfortably furnished. There was a television set, a cassette player and a radio-telephone. There was a rug on the floor in the living room and a very simple couch. One electric socket provided light and a hot plate served for cooking requirements. Water for drinking, cooking and bathing was available at the community pump house. The entire St. Pierre family spoke in their native language which was very interesting to listen to.

Later that evening Darren, a 16 year old friend of the family, came to the house and we set out on foot to see the town. Our short hike took us past the RCMP office, several homes, a dock and the airport. We also observed a diesel electricity generating plant and the local general store which had gasoline pumps. The price of a litre of regular gasoline was 62 cents. Before leaving Edmonton, we had filled up our tank at 36.5 cents per litre.

When we returned to the St. Pierres' home we found that the light and television were still on. The house was serving as a meeting place for those interested in the blockade and everyone there was happy and talkative. I was given a bed to sleep in. I very much appreciated this gesture as I knew that at least eight or nine people slept elsewhere in the small house.

Early the next morning I boated to the local general store with

236

Mr. St. Pierre's son. It contained a very wide selection of inventory, including food, clothing, hardware and tools as well as merchandise related to trapping, which I found very interesting indeed. After finishing breakfast at the St. Pierres' I repaired the front steps of their home and then accompanied Mr. St. Pierre to look at his fishing nets. We went southwest on the lake several kilometres from his house. There we found a lot of fish in his 7.5 centimetre mesh nets. He uses the fish to feed his family as well as his sled dogs.

By the time we returned to the house several neighbors had arrived. They were interested in attending the blockade and consequently we loaded up the boat and were gone from the reserve before lunch. The crossing was quite uneventful but we noticed several thin rain clouds on the horizon. We were watching their steady movement northwest across the lake when we saw a large cloud of pollution enter the atmosphere. My friends explained to me that the pollution was caused by the uranium mining activity.

When we arrived at the dock near the campground we were astonished to find 16 boats anchored there. Natives from several other communities had arrived to lend their support to the blockade. By that time a regular shuttle service had been established between the campsite and the blockade.

It was raining by the time we reached the blockade. What a transformation had taken place! A field kitchen had been set up and there were toilets and many tents. Several loads of food had arrived and an endless chain of vehicles stretched along the road from the mine gate.

One of the men at the demonstration had a pair of binoculars, which we used to observe the activity on the other side of the fence. We saw a 1/2 ton truck with a coffee machine set up on its tailgate. There were several officials from the mine with binoculars who were watching us watching them. We were about 200 metres apart. All of the men were about 50 years old, and each one wore a concerned look on his face and a white hardhat on his head.

The next day, June 16, Sid and I started the long drive south. When we arrived in La Ronge the following morning at 8 o'clock, we spotted two RCMP paddywagons heading north to the demonstration. We called the Lac La Hache Band office to warn them. Next we stopped in at the La Ronge newspaper, *The Northerner*, where I was interviewed by the publisher and editor, and an article appeared on the front page in the next issue, June 19th. Then we headed home to Edmonton.

Discussing the RCMP ultimatum.

June 17, 1985:

The mood changed again, this time for the worse. The RCMP gave the word that Babcock would not meet until after the blockade was lifted, and the place would be either the mine site or Eldorado headquarters in Saskatoon. Only a small delegation of Dene Chiefs and elders would be permitted to attend. And to top it off, the RCMP said, "The trucks will get through," and that people would be arrested if they did not move off the road.

It was later leaked by an Eldorado employee that Babcock had been willing to meet with everyone at the mine gate but that the RCMP had intervened, perhaps with orders from Ottawa. A confrontation was being forced. It was time to make a decision.

In Saskatoon about 35 people held a short support demonstration in front of the Eldorado office. And in Ottawa, Bill Blaikie, MP, spoke out in the House of Commons against uranium mining and in support of the blockade. Part of his statement follows,

238

Two RCMP Officers watched as the final decision was being made to leave or stay.

reprinted from the official House of Commons Debates report.

Early in the afternoon the Wollaston people and supporters sat in a circle amongst the trees and held council. Meanwhile, unknown to the blockaders, a convoy of trucks had left La Ronge and was on its way to the mine. After a few hours discussion, with the RCMP looking on, the decision was made to lift the blockade pending the outcome of the meeting with Eldorado.

By 7:30 in the evening all the blockaders had left the gate. In a couple of easy swipes a large earth moving machine cleared all the rocks and trees off the road. An hour and a half later 22 eighteen-wheeler trucks rolled into the mine site, including some SINCO trucks. Never before had such a convoy travelled up the long northern road. This disproved claims by industry spokespeople that the blockade had had no effect on mine operations.

CANADA

Monday, June 17, 1985

OFFICIAL REPORT
(HANSARD)

House of Commons Debates

VOLUME 128 • NUMBER 126 • 1st SESSION • 33rd PARLIAMENT

Mr. Bill Blaikie (Winnipeg-Birds Hill):
Members will be aware, I hope, that at this very moment, and since Friday, there has been a blockade going on at Wollaston Lake in northern Saskatchewan having to do with the concern of people there about the extent of uranium mining going on there and its potential expansion. I support that group and I feel for their desperation in trying to get someone to listen. I hope the Minister of the Environment will do the right thing and initiate an environmental review process, which I understand has never really taken place up there, either provincially or federally, regarding the new mines being planned. I hope she will try to initiate a process whereby the people there could be listened to and the effects of these new mines on the renewable resources economy of that area, on the native community and on the environment in general, could be assessed.

To be honest about my own biases in this area, I say that in the context of my opposition to uranium mining anyway. I put that opposition on the record when I first came to this place, even when there was a NDP Government in power in Saskatchewan. My opposition has been consistent throughout, and so when I am critical of the provincial Government of the day in Saskatchewan for not doing what I think it should do, I want no one to think that I am singling out the Progressive Conservative Government. I think uranium mining is a mistake no matter who is responsible for it and no matter to which

Bill Blaikie, Member of Parliament, New Democratic Party.

241

Government the Crown corporation involved is responsible. It is one of the most hideous way we have yet developed as a white industrial society of trampling on the lives of native communities. We do this in the name of the international arms race and sell uranium to countries like France so they can blow up innocent little islands in the Pacific and continue their ecological crimes, sins against the planet which they seem determined to commit regardless of the opinion of the international community, or anyone else for that matter.

I think it is high time that Canada got out of the uranium mining business altogether and said that we do not want to play a part in this. We do not want to be a part of the international uranium system which is part and parcel of the nuclear arms race. We have often got up in this place and been critical about the arms race, yet at the same time we remain silent about that aspect, the uranium mining which is taking place in northern Saskatchewan.

In that context I would call on the Minister of Energy, Mines and Resources (Miss Carney) to institute the kind of public or parliamentary inquiry into the whole nuclear fuel cycle which was promised by the Progressive Conservative Party in 1979. That promise was in its Throne Speech of 1979, but since it came back to power in 1984 we have heard absolutely nothing about such an inquiry. I used to have the impression that that Party was critical of the nuclear future which was imposed on Canada by the Liberals without any real consultation with the public and without any real debate about the role that nuclear power should have in our energy future. I had some hope that that was one area in which the Conservatives might do something I would agree with, but we have had complete silence since the election of September, 1984. When the silence has been broken, there have been statements from the Minister of Energy, Mines and Resources which are, to put it bluntly, quite frightening. The Minister has said that the only problem with Candu and the whole nuclear power option is a marketing one. That is not a view that goes to the root of the more fundamental question of whether or not this is an energy option that we ought to be pursuing in any event.

Greg R. Land, 1982

Wollaston

Blood red this land now
Reflected in cores
Shipped to the lab, analyzed
How often they need to be told
It's a strength they can never control
Blood of the ancients
Reflected in sorrow
In faces grown old with this land
The rich come to take what belongs to no one
They lay claim with man's white hand
They lay waste to Indian land

In northern Saskatchewan, near Wollaston Lake
Compatriots struggle to stop those who rape
Arm in arm cross the road they cry out you will not pass
But twenty-two armed trucks come up the long road
Three hundred miles don't seem far today
Brothers and sisters move off of the road
Let villains and guns have their way
It seems bullets will always hold sway
Death rock and guns they kill on the roadside
In Latin America or Wollaston Lake
It's an ancient sad story, a fight for gold glory
The price is our lives and our pain
How long do we plan to pay
How much do we plan to pay

Latin American children and babies
are thrown in the river and float out to sea
These are the same children who die by the waters
poisoned by villains
in our true north strong and free
Some may soon die now, mothers and fathers
partners and lovers, your babies and mine

They'll come for your babies and mine
In this fight do you stand by my side

We too are the core now
and soon they must know and
Shake in the wake of our wave
How long before we all show

244

We're the strength they will never control
The price they demand is too high
No more gain from our lives and our pain

Their tactics will stale as we witches wail
We could bring down their night in a day

Away ghostly shadows they cannot lay claim to
They'll tremble at our very sight
We be witches that wail in the light
We be witches that wail in the light

Death rock and guns they kill by the roadside
In Central America or Wollaston Lake
It's an ancient sad story, a fight for gold glory
The price is our lives and our pain
How much do we plan to pay
How long do we plan to pay
Just how much do we plan to pay?

**– A song by Penny Ruvinsky (Vancouver, B.C.)
written with guitar music, summer, 1985.**

GLScrimshaw
c85

The Outcome

The June 1985 gathering and blockade focused attention on the Wollaston region problem, but what about conditions in the native communities and influence on the uranium industry? The answer is clear. Conditions in the communities are getting worse and the uranium industry continues to expand. On the positive side, a network of concerned people was formed, both in the north and south and between the north and south.

An outcome of the blockade, before the event even took place, was circulation of a number of false rumors in Wollaston. For example, people became worried that if the protest continued welfare payments would get cut off. Further, there was talk of the road having to close if mining stopped. One of the most vicious rumors was that the southern protesters were all "Greenpeacers" who were against hunting and trapping and particularly against the caribou hunt. It was clear where this last rumor originated – from Jean Megret, Sol Sanderson and Eldorado PR men, who used it as a divide and conquer tactic.

As far as welfare payments getting cut off, the government would really have a war on their hands if they tried that. The controversy would be so great that they could never get away with it. However, welfare benefits to Wollaston were being slowed down before and after the blockade. Three good examples of this are the slow progress on new housing, and the fact that construction of the new school and fire station almost came to a halt. New houses were built in Wollaston the summer of 1985, but an entirely inadequate number. In the case of the school, the plans had been finalized and a site cleared in 1984 and that's where it stopped.

The fire station was almost completed in the winter of 1984–85 and a fire truck was expected on the first barge in late May 1985. In Wollaston preparations were made for the fire truck's arrival, including gathering names of people to join the volunteer fire brigade. Suddenly, just weeks before the truck's scheduled arrival the federal Department of Indian and Northern Affairs (DIAND) cancelled the whole program. The reason given was that there wasn't enough money. Instead, the barge hauled in building supplies for a new RCMP house and office.

Whether the reason for the slowdown in welfare benefits was an inefficient bureaucracy or a conscious attempt to subdue the people, the effect was the same: demoralization and the feeling of

246

powerlessness. The limited housing and jobs available are distributed in such a way that people are forced to compete with each other. This is an ideal situation for the mining companies because instead of protesting, people fight amongst themselves and turn against the Chief and Band Councillors for failing.

The June 20, 1985 Meeting And Press Conference

The sought-after community meeting in Wollaston with Eldorado officials did not take place. Rather, on June 20, 1985, three days after the blockade was lifted, there was a private meeting in Saskatoon at the Bessborough luxury hotel. Those in attendance included the Chief and Council of the Lac La Hache Band, Sol Sanderson and others from the Federation of Saskatchewan Indian Nations (FSIN); and representatives of Eldorado Nuclear Ltd., the provincial and federal governments. Spokespersons from anti-uranium groups were specifically excluded.

The meeting resulted in only a minor variation from the status quo. Two committees were formed: a "communications committee" to increase communication between Wollaston people and Eldorado Nuclear Ltd., and an "economic committee" to determine how Wollaston people could benefit from mining. No "close the mine committee" was created, nor did the economic and communications committees achieve their stated objectives.

A turning point was reached however, in March 1986. At that time it was announced that a 36 km, 25,000 volt power line would be built to Wollaston from the Rabbit Lake mine. This is an extension of a new $50CDN million transmission line from existing generating stations near Uranium City to the Rabbit Lake mine. The project was made feasible by Eldorado Nuclear Ltd. agreeing to use 50 million kilowatt hours of electricity annually until 1996. According to the Saskatchewan government, the powerline to Wollaston was not in the original plans, but was made possible when Eldorado agreed to pay 50% of the costs.[6]

The blockade did not slow expansion of the uranium industry. Two main reasons for this failure are the relative powerlessness of the Wollaston people, and the pro-uranium mining policy of the FSIN. Sol Sanderson, an ex-RCMP Officer and president of the FSIN, expressed his pro-nuclear position to the media on several occasions. This suited Eldorado Nuclear and the government perfectly, but weakened Sol Sanderson's support at the

reserve level.

Sanderson was the dominant speaker at the press conference held by the native leaders after the June 20th meeting. He made false and outrageous statements in an effort to split the Dene from outside supporters and discredit the support work that had been done. Chief Kkailther, participating in his first press conference, seemed intimidated by Sanderson's brazen paternalism and implied economic catastrophe should the mining stop. Sanderson downplayed the importance of the blockade and advocated stronger Indian involvement in the uranium industry.

Following are excerpts transcribed from a tape recording of the June 20th press conference. All the enlarged statements were made by Sol Sanderson. The next day, June 21, MP Jim Manly, Indian Affairs Critic for the New Democratic Party, made an,

248

anti-uranium mining statement in the House of Commons debates. His statement is reprinted after the transcript. Concluding the chapter is a letter from the Saskatchewan Association of Northern Local Governments (SANLG), and a statement by anti-uranium mining advocate George Smith, Mayor of Pinehouse and Chairman of SANLG. The annual budget of SANLG is about $5,000CDN, while that of FSIN is about $20CDN million.

The June 20, 1985 Press Conference

Sol Sanderson (SS):
Government representatives and the Chiefs of the Athabasca region addressed a number of the specific concerns that the Chiefs have in respect to uranium development and northern development. We also have with us Milt Burns, District Representative from Prince Albert District and Chief Roy Bird who is on the Chief's planning committee for the north.

In the northern development strategy, we're looking at the uranium strategies in specific terms today and in the previous session. We've looked at the forestry development and hydro two weeks ago in Saskatoon here with the Athabasca Chiefs and the uranium hydro access to the Bands in northern Saskatchewan and to the industrial developments in the north. The Chiefs have agreed with that approach.

Just to review with you some of the background in terms of law, under the treaties and the constitution, as you know we have continued right of access to resources both on and off reserves, protected in the treaties and the constitution. We're going to quickly highlight this: the funding to date we're uncertain, but we know that between 1974 and 1982 there was $159 million of federal funds spent in northern Saskatchewan, for what we're not sure. But we know it did not impact on the reserves and the Bands in any way directly for improving conditions, so that's what the Chiefs are concerned about. Provincial government expenditures to date, industrial expenditures we're not sure of. We want to find all that out. From exploitation of resources to date, there are no returns and no benefits, except for some employment opportunities, but we need much more than that in terms of evidence of returns to our people in the north.

An example of industry and governments, non-Indian governments co-operating to benefit the people in the communities from resource development is the James Bay hydro development.

There's the Manitoba Flood Agreement which is a current one and a substantial one in the terms of returns to the communities and to the people of the reserve and the Bands. There are the plans for the Prince Albert District Chiefs that they have in place for looking at the developments in terms of uranium and so on.

The Chiefs have done a major study here on uranium development and we can give you copies of the executive summaries of the recommendations.[7] I met two weeks ago with the Athabasca Chiefs. They had a two day session on this economic strategy for their areas, and they asked us to get involved and co-ordinate that so we can get on with the developments formally. As a result of that meeting we have organized a session today. Along with that we also addressed some of the concerns that the Chiefs have to deal with in terms of the protest in the north, but that was not the primary reason for the meetings here.

The returns from northern development and respect for rights are two of the areas we want to focus on. In spite of all the developments in the north and the money spent, the symptoms are still there: poverty, high unemployment, high suicide rates, high infant death, alcoholism. In the process we have loss of access to traditional territories for hunting, fishing and trapping; and if you have a look at the Manitoba Agreement and the James Bay Agreement, there is respect for continued right of access to those areas.

"We are going to form an economic committee."

What we've agreed on this afternoon in the meetings including the one we just concluded with everybody: industry, federal and provincial government representatives and ourselves, the Chiefs of the Athabasca region and the District Chief of the FSIN, we are going to form an economic committee. The terms of reference of the committee will be struck, broad terms of reference that include the objectives impacting on the reserves, Band developments, and federal and provincial government and industrial development objectives. The more specific work plan will be focusing on the reserve and Band and Indian development as it applies to resources and lands and waters both on and off reserves; mill impact on the water quality in terms of potable water and access to quality water. We'll also deal with the resources needed to provide improved conditions on community infrastructure for sewer and water systems, roads, electrification, and different areas like that. The strategy will include looking at wildlife resource policy. In

250

SINCO (Saskatchewan Indian Nations Corporation) is owned by the FSIN. SINCO Security guards the Cluff and Key Lake mines.

the Manitoba Agreement the residents have first right of access to all resources in areas of wildlife and so on, and it leaves to the provincial government to exclude non-residents and people from outside the area coming in and accessing those resources. Where there is industry moving in to a trapline area, that land has to be accessible elsewhere and money has to be affordable for trapline development and so on.

Concerning environmental impact policy, we want to know what the immediate and long range impact of uranium development's going to be, and we're going to get involved in setting some standards. I pointed out to both industry and government officials today that in my travels to different parts of the world, I have seen non-Indians go to great lengths to build 16 and 14 retainer walls to protect themselves

251

from each other. What we're saying is that if waste products of industrial developments are so serious in respect to its impact on environment, then we should be prepared to go to 14 to 16 retainer walls if we have to instead of the 2 or 3 that are now being looked at.[8]

"There is a serious lack of communication...to the point where some of the protesters from the outside in the north right now are protesting to stop uranium period."

The liaison committee is the second committee we will strike as a communications committee with the co-chairs of the economic committee involved and one or two people from each of the Bands. There is a serious lack of communication about developments generally, not just uranium, to the point where some of the protesters from the outside in the north right now are protesting to stop uranium period. These same protesters up there right now are the same ones involved as well in trying to stop the leg hold trapping. So they are involved as well in trying to stop the traditional livelihood of the people that the Chiefs represent. And they are also saying that we cannot get involved in contemporary type economic opportunities. So we can not sit by and let people run

---

"Chief Robillard insists that we put on the agenda as well, mining and that we are going to control mining...a number of Bands are now selecting potential mining sites so that we can get on with the control of mining."

---

over us like that, whether they are pro-developers or anti-developers. We are going to make a choice about how things are going to move ahead.

Chief Robillard insists that we put on the agenda as well, mining and that we are going to control mining. We are not looking at just the contracts and the employment opportunities, but a number of Bands are now selecting potential mining sites so that we can get on with the control of mining.

So ladies and gentlemen that's about it for highlights this after-

252

noon and where we're headed. We have agreed with the Chief here that in Wollaston we will go up to a Band meeting, we will go up to Stony Rapids with the Chief and Council to a Band meeting there. The same with the other Chief who is not here today, Chief Mercredi of Fond du Lac. We will go up there and meet with them. Eldorado and the Chief and Council from Wollaston will get into immediate discussions next week. There has been a breakdown in communications between the Band and the company, so they will deal with that very openly and frankly amongst themselves next week. But we will be going up to the Bands and workshop with the Band generally. So if you have any questions feel free to ask them, ask the Chiefs any of the questions you may want to put to them. We want to be able to set the record straight too because we have had the media printing messages from everybody except the representatives of the people in the north, and that's the Chiefs we have here. So if you want anything in terms of information about the developments up there and what's expected of those developments, then the Chiefs are here to answer as well.

Reporter: This committee you formed which the Band members will sit on, the companies will sit on it as well?

SS: Yes, federal and provincial governments.

Reporter: Why all of a sudden now? Have these negotiations been going on for a long time?

SS: We have been involved in the planning some time ourselves with our chiefs and the District Council, and the FSIN economic

---

"We do not want anyone exempting themselves when it comes to the bottom line of accessing money towards Indian development."

---

commission and industries only felt the need to come forward now I guess, because of some outside pressures, outside elements and that some pressures were there a year ago when there was some lobbying being done in Europe. We told industry and governments alike that we want to sit down and discuss it formally but we do not want anyone exempting themselves when it comes to the bottom line of accessing money towards Indian development. At the Band and community we want to know what the benefits are going

253

to be in real terms to Indian people. So that's the stage it is at, at this moment. We are not going to be preoccupied with protesters because those same protesters from the outside, like I say, are also protesting in Europe and across the country now with Greenpeace opposing our traditional livelihood in terms of fur, hunting, trapping, that includes the Wildlife Federation. They might be wise to get on our side pretty quick.

"We are hoping that the economic committee will come out with a comprehensive agreement."

Reporter: Could you just sum-up what you hope this committee you developed this afternoon will achieve or is working towards?

SS: We are hoping that the economic committee will come out with a comprehensive agreement that will direct fiscal resources to community development and benefits for our people in the community by Band, and the Bands collectively in the north.

Reporter: And this is from uranium mines in the north?

SS: That's the focus today. Previous sessions we have had were on forestry, and we are also having a series of sessions right now on trapping, hunting, fishing and gathering.

Reporter: Does this mean that the Chiefs of northern Saskatchewan are prepared to accept uranium development on northern Saskatchewan?

SS: Do you want it from the Chiefs or do you want it from me?

Hector Kkailther (HK): For that it depends on my Band members. I've been sent down and now have a few questions for them, so I'll wait till I get back to them and then I'll answer you.

SS: Did you want him to outline his expectations from the development?

HK: Well, it's pretty hard for me to say all these things here right now, but these things should be done in the first step. We have to outline all these items here, that's a good step for me. As for

254

SINCO Trucking hauls uranium to Saskatoon from the three operating mines. Some SINCO trucks were in the 22 truck convoy going into the mine right after the blockade was lifted.

myself, that's what I'm for, I'm for employment. I'm not Chief for fighting with government or protesters or something like that or whatever, so as I stand now depends on my Band members.

Reporter: There have been letters coming from the Lac La Hache Band which say you want uranium mining in northern Saskatchewan to stop you don't want Collin's Bay B-zone to go ahead. Do you still hold firmly to that position or are you prepared to see what could come out of the work of this committee?

HK: Well, like I say, it depends on my Band members right now. What I told them last couple days, and we're having meetings last three days in my Band this whole thing. So I broke down the blockade last couple days for this meeting right now. So I hope they understand and I hope it runs well from there.

Reporter: Are you going to be suggesting to the Band that they hold off on the blockade for the time being or are you going to set the blockade up again?

HK: That depends on them.

Reporter: OK, when will they make a decision do you know?

HK: Well, its pretty hard to say, ask these questions right away, like I say, it depends on my Band members there.

Reporter: OK, we've heard that the blockade might come up again tonight. Do you think it will be back up again or not?

HK: Well, they're waiting for my word right now so I hope I get back tonight and explain all these things and I hope they understand.

Reporter: What will you recommend, that it goes up or not?

HK: Well, like I say it depends on my Band members.

Reporter: What do you think? Obviously you agree to this idea of the economic committee being established, so?

HK: Yes, I agree with idea about this thing but these things should have been explained to the Band members before. The first time they started mining my local people didn't want mines to operate in all the area. That's about 12–15 years now. Local people didn't get recognized or listened to.

Reporter: I get the impression though, just from hearing Mr. Sanderson and now you, that if the committee agrees to do what you, you know if you negotiate some certain things as far as trapping and fishing and access for example, you wouldn't mind uranium development if as long as those things were agreed to, is that a fair assumption?

HK: It's a fair assumption for myself, but it depends on my Band members.

SS: We want to deal with the environmental protection matters as well, as part of the whole development strategy. We don't want, like I say we're not in the business to support pro-developers at the expense of everything, there's a limit to everything, not just uranium development. I don't know if you got the Chief's point, but he was trying to express to you that both industry and governments have ignored them and have no respect for their Chief and Council's direction and position on this point. That's what he was trying to explain to you at the end of his comments there, and what he said is all this should have been in place before they moved ahead with the developments.

Reporter: If the working committee provides some satisfactory, I guess revenues that might come in, are they prepared to see uranium development go ahead, you know of course keeping into account the environmental considerations as well?

Chief Roy Bird: OK, what I am here for is, I was not involved in that protest, but I understand what Hector's trying to do, he said, "We are not trying to close the mine but for the long run you see that you take care of that mine real good and keep caution of what they are doing in the long run, it would hurt our environment and wildlife. And I would like to see these mining companies come to our community and tell the people what they are doing at Rabbit Lake, we have not had that for a long time." I told them today, tell me where this thing all started?

"We have to talk about dollars and cents, that's the bottom line, that's why we're here."

Reporter: Have the mining companies agreed to what you're talking to us about this afternoon?

SS: All except one so far and they're going to get back with us tomorrow morning.

Reporter: Can you mention names?

SS: Well I think it's the one that stands the most to gain, I will put it at that, like I told them but I do not see any of the companies not wanting to get involved.

Reporter: It would mean the transfer of some of their revenues to the Indian Bands is that what we are talking about here?

SS: We are talking about the bottom line being dollars, yes, for Indian development, for Indian control. And if it means accessing some of the royalties and taxes before they go to the provinces and the federal government, then that is what it means. That's simple as that, we have to talk about dollars and cents, that's the bottom line, that's why we're here.

Reporter: So you're going to be agreeing to negotiate that principle of sharing some of the resource revenue?

SS: Well the stage we are at today is we had three hours to work at it, and it's basically agreed to in principle by all parties, I would say. We have to come back to another meeting, we will be sending out a letter, when we will call this initial session to the participants today and formalizing commitments to the economic development committee and getting appointments made to it. Then we will sit down and prepare initial draft terms of reference for the work to be done, and we will have to go up to the Bands and discuss with the Chief and Council the approach. We expect we should be able to get something done in weeks, not months or years because of the precedent set by James Bay and Manitoba Flood Agreements and so on.

Reporter: Why was this not started before?

SS: Why was this not started before? I think the Chief put his finger right on it just a moment ago. We did not have the respect

---

"They even want us to address the protesters. Well, that's some of their people, that is not all the Chief's people up there, that is some of your kind up there that are lacking communications."

---

from both levels of the government nor industry to take our direction and our leadership on these developments on these approaches, and that's been sadly lacking up to now. Now they're looking to us because they are in trouble. They cannot find the answers, and we have said we have had them for some time. They even want us to address the protesters. Well, that's some of their people, that is not all the Chief's people up there, that is some of your kind up there that are lacking communications. You give them a lot of press by the way, I think you give them too damn much, because they are also trying to cut off our traditional livelihood. And some of the people you're going to do not have status in the community for leadership.

Reporter: You're angry at us?

SS: I am angry at this moment with you people, yes, for that because I do not appreciate you going to people off the streets and dealing with them in terms of their position on issues, not when we got the spokesmen ourselves in place to represent us properly. That is all I can say, that's as far as I will go on it.

Reporter: Just on that point about the protest up there, do you feel that was engineered?

SS: Orchestrated.

Reporter: Orchestrated?

SS: Well, it's orchestrated by the people you publicize, Goldstick and Graham. They were in Europe representing our people in Saskatchewan, in northern Saskatchewan and all over the place. Said they were the Chiefs of Canada from Saskatchewan representing thousands of people from Saskatchewan and saying they spoke for us. Saying that we oppose uranium development and all that stuff. At the same time they were over there lobbying against the hunting, fishing, trapping rights of Indian people.[9] So make up your minds who you are going to expose in terms of what needs to be said in an honest way.

Reporter: But they claim they're speaking for the people of northern Saskatchewan.

SS: I just finished saying we have the people elected to speak for Indian people, and let's have the same respect from the media as we demand from the governments and industries up there.

Reporter: Have any figures been batted around? How much money are you talking about?

SS: Well the Manitoba Flood Agreement that you're talking about in James Bay is sixty million dollars a year towards the communities and the James Bay region. That's fifty-five hundred people, so many communities. I know that I was involved in a part of that

259

exercise up there with the James Bay Cree and in Manitoba, I am not sure but you can check it out very quickly what it amounts to. I understand they have some fifty-seven million or fifty-four million dollars available just for community infrastructure.

Reporter: Has any figure been mentioned in Saskatchewan?

SS: For Saskatchewan, we will work with the committee structure and we will talk with the people, what their priorities are and how much it is gonna cost and yes we will come up with the figures.

"We expect industry to go on and developments to move along."

Reporter: Have you got a bottom line?

SS: No, we expect industry to go on and developments to move along so we do not have a specific bottom line at the moment. We know some of the requirements and demands are there and we know there are existing resources that could be co-ordinated for that, facilitate that.

Reporter: Have the companies mentioned a maximum?

SS: No, we do not have a maximum or minimum at this stage. We want it to be an orderly process in place so that we can access those resources to those developments. Like we said, we have had experiences with federal and provincial agreements in the past, the Northwestern Northlands Agreement, where they accessed $154 million between 1972 and 1984 for Indian development, community development and so on. None of it got there. It went somewhere in the north but none of it made it to the communities. Now we have a Northern Economic Agreement, we have other agreements in place, federal, provincial, but the Chiefs are not feeling the impact at all nor are the people. What we're saying is we want a specific fiscal agreement, then we will focus on directing those resources there.

Reporter: What's the government's role in all this?

SS: Well, we met with Indian Affairs as well and we told them they have the lead role in this situation federally, because of their legal obligations under the law and that both federal and provincial governments have a legal responsibility as well when it comes to rights and resources. So we want to be able to co-ordinate all that through the economic committee work and get something focused on each of those areas we outlined.

Reporter: But what if you set up an agreement, let's say with Eldorado, and in a year down the road and they say, "Sorry, our profit does not look good this year."

"If uranium wants to flourish and develop, they need our help, they need our support."

SS: Well that's the nature of development, if they are going bankrupt, then everybody goes bankrupt. That's our point, if uranium wants to flourish and develop, they need our help, they need our support. If it is not there then they lose too. And if the company's not profitable, nobody benefits. That's a clear message to all of us. But where they are making profits we want a share of it.

Reporter: Is this the new community system now, for every new mine that comes on stream, even a gold mine for example?

SS: Yes, the Chiefs started with forestry. I do not know how many meetings they had with forestry.

"Where they are making profits we want a share of it."

Reporter: Can you identify who is on the committee?

SS: Not at this moment.

Reporter: In terms of which department of government and companies?

SS: No, we did not get to that stage. There is the mining company association, uranium mining association. I imagine they would want representatives, but leave it to them to appoint these representatives.

Reporter: Are you looking at cash flow?

SS: The cash? We are looking for something immediate, yes, definitely. That is why the Chiefs are here.

June 21, 1985

OFFICIAL REPORT
(HANSARD)

CANADA

ᕼouse of Commons Debates

VOLUME 128 • NUMBER 130 • 1st SESSION • 33rd PARLIAMENT

INDIAN AFFAIRS

LAC LA HACHE INDIAN BAND - EFFECT OF URANIUM MINING

Mr. Jim Manly (Cowichan-Malahat-The Islands): Mr. Speaker, in the last few days Indian people from the Lac la Hache Indian Band, and environmentalists, have blockaded the road to the uranium mine at Wollaston Lake operated by Eldorado Nuclear Ltd.

The uranium deposits at Wollaston Lake are among the largest and richest in the world, and Eldorado Nuclear plans to keep mining there for at least 20 years. Then Eldorado will leave, but it will not take its garbage with it. Indian people will have to live with radioactive pollution for generations and generations.

Indian people take very seriously not only their aboriginal rights but also their aboriginal responsibilities to take care of the land for future generations.

As inhabitants of planet Earth, we desperately need to realize that the struggle of the Lac la Hache Indians is also our struggle. We too have a responsibility to preserve the earth for future generations. The mining and the pollution must stop.

Statement By Pinehouse Mayor George Smith, August 7, 1985[10]

We're against the uranium mines for sure, but just how are we gonna stop them? That's the problem.

No matter how hard you try to explain to the government that uranium mines are not doing any good for us and not employing our people and not giving us any benefits, they always make it sound good in some way, like they'll give jobs. But that's just a line of bullshit because we don't get the jobs or anything out of it. Most of that money is going out to the south and very little is coming back to the north. The mining companies just care for themselves. That's all. They don't give a shit about the people. They just try to pretend there's jobs for us.

They have all the power. They said the road block was caused by outsiders and not the residents of Wollaston Lake. I know damn well it started with the people of Wollaston Lake because I was in Wollaston at the end of April and there was 250 people in their Hall and they were saying they don't want uranium mines. They never did want uranium mines but it went ahead anyway. They feel powerless. Their leaders and their Chiefs all have to get together and make up their minds whether they're all going to be against uranium mines or else for it. The Chiefs get trucking and other things so they're pretty quiet about it.

The arbor made by the blockaders at the mine gate. The photo was taken a few days before mine staff knocked it down in late July 1985.

263

c/o GENERAL DELIVERY, PINEHOUSE, SASK.
S0J 2B0
Phone 884-2030

11 July 1985.

Sid Dutchak,
Minister Responsible for SMDC.
Legislative Building,
Regina, Sask.

Dear Mr. Dutchak,

At a meeting of the Saskatchewan Association of Northern Local
Governments on June 26th, 1985, the Mayors and Councillors of
many northern communities discussed the uranium developments
proposed and under construction in the Wollaston Lake region.

We continue to support the community of Wollaston. The following
motion was unanimously passed.

> "SANLG is opposed to uranium mining because we get no
> benefits from the mines and because of their after-effects
> on the health and environment of people of the north."

As minister responsible for SMDC you must begin to recognize
the position of northern people on this issue.

Sincerely,

George Smith,
Chairperson.

c.c. Grant Devine,
     Neil Hardy,
     Hector Kkailther,
     Jonas Hansen

264

A Critique

This critique is limited in scope to only a few aspects of the gathering and blockade.

Why was Eldorado Nuclear Ltd. able to triumph so easily? An important part of the answer is the opposition's lack of resources. This is especially true for the native people, but also for the non-native anti-uranium movement. There is not a single person in Saskatchewan, or Canada, who has a salary to work full time against uranium mining. Considering that there is more uranium mining activity in Canada than any other place in the western world, this is a sorry state of affairs.

With regards to the blockade action itself, lack of preparation for arrest was an important factor. Dean McKay from the Montreal Alliance For Non-violent Action wrote after the blockade,

If there was an arrest situation the lack of preparation could have created a feeling of disempowerment when this should be empowering. If, in fact, one of the reasons for discontinuing the blockade was fear that people weren't prepared, it's clear that lack of preparation can make us defeat ourselves. I'm not saying that the discontinuation was a defeat (I don't know). I'm using it as an example to illustrate the importance of organization.

The following essay by Jack Ross discusses a number of aspects of non-violence in more detail.

In a broader context, the issue of uranium mining cannot be separated from the issue of how Indian people are treated in Canada. A stop to uranium mining alone is not the solution to ending poverty in the north. Humane and environmentally sound forms of economic survival need to be implemented. More support must be given to community based Indian land claims and development projects. The term "community based" is used to distinguish from neo-colonialist activities of groups such as SINCO.

Non-native Canadians especially have to realize that the Canadian government treatment of Indian people is inhumane. No matter how good and honest the intentions are of a Chief and Council, they simply cannot make significant progress under the current arrangement. One of the problems is that traditionally, the extended family or groups of families choose representatives to participate in decision making that involves a larger group.

265

But, the only authority the government recognizes on a reserve is a Chief and Council elected by the government's rules. The election system has been imposed from the outside without participation of the people effected.

The biggest barrier to improving reserve conditions is that the Indian peoples' needs are decided by urban based government bureaucrats rather than at the community level by the people themselves. The federal and provincial governments make all major decisions, such as those dealing with education and local resource use, including mining. There is rarely any encouragement of community input.

A good illustration of this point is an exchange that took place at a meeting on April 11, 1985 between the Lac La Hache Band administration and the federal Department of Indian Affairs and Northern Development (DIAND). The major point of discussion was the new Band budget. A DIAND official presented a detailed, already determined budget. Chief Kkailther responded by stating that the Band members must be involved in planning it. The highest ranking representative from Indian Affairs responded calmly and coldly that, "There are few areas open for negotiation."

Gandhi At Collin's Bay

By Jack Ross, October 30, 1985

The blockade on the road to the Collin's Bay mine was referred to in pre-event publicity as nonviolent. In actuality no physical violence occurred in addition to that which the mines perpetrate on a daily basis with their radioactive wastes and environmental destruction. There were specific prohibitions of alcohol, drugs and firearms, and these were observed. There was some discussion of nonviolence during training sessions and role playing of potentially violent situations. Beyond this, the nature of nonviolence, as philosophy or theory received little attention.

It was evident in our behavior and in our expressed thoughts in our final evaluation that there was considerable disagreement about goals and methods. We dealt with these more or less successfully as they occurred. Certainly the situation that we found ourselves in was not conducive to long discussions about philosophy. We dealt with immediate problems and not much with abstractions or principles.

Now that the event is over it is appropriate to take a longer look at our actions in light of the nature of nonviolence. My goal is not to compare what we did or did not do with some ideal. I think that would not be constructive. Actual human behavior always falls short of the ideal, especially when looking back. Rather, as a way to begin a dialogue about protests and blockades of the future, I shall use a review of the principles that Mohandas Gandhi came to see in 1930 as central after 40 years of experience. There will be some casual evaluation, but the main intent is to stimulate analysis.

Gandhi himself viewed his understanding of nonviolence as incomplete. He titled his autobiography "The Story of My Experiments With Truth". He never wrote a final theory but continued his explorations until the end. Nevertheless, certain topics kept coming up throughout his amazing career. I have selected one version of principles that he requested as vows of his most devoted followers, the members of his ashram (community farm). This then is something of a statement of the requirements for perfection. Do we want to make that sort of demand of ourselves? Gandhi asked for this perfection from many but accepted less in practice. Where do we fit in this range of possibilities?

267

*Jack Ross, peace activist and retired sociology professor, partici-
pated in the blockade and the SAND tour.*

*"Outside of some practical matters concerning organization, I am
mostly concerned about the nature of nonviolence and the tactics
that are called for in an unbalanced bicultural situation."*

– Jack Ross, June 24, 1985.

These principles are: Truth, "ahimsa," chastity, and nonpossession. These four were selected by the editor of "Nonviolent
Resistance," B. Kumarappa from a longer list by Gandhi that
includes fearlessness, palate control, nonstealing, bread labor,
equality of religions, anti-untouchability, and "swadeshi" (making and using local goods). These others would be instructive
too, but I think the most important issues are raised by consider-

ing only the first four. As well, I will make a comment on "swadeshi."

Truth

This was a basic concept for Gandhi. He sometimes expressed it in religious terms: God is Truth, and Truth is God. The individual who practices "satyagraha" (i.e. truth-force) must have a wholehearted and selfless devotion to truth in thought, speech and action. Gandhi did not draw back from this conviction when faced with the fact that individuals define and pursue God and Truth in different ways (or that they pay attention to neither). His view was that those who take the wrong path eventually stumble. "God as Truth has been for me a treasure beyond price..."

I think that our experience at Wollaston exhibited a considerable implicit agreement with the principles of Truth and a desire for consistency of principle and action. Perhaps it is inherent in a group assembled for a single event from different places and different cultures that not much effort is put into exploration of abstractions like Truth or God, or alternatives. A close look at the way Gandhi trained participants as diverse as Muslims, Hindus, Sihks, Pathans and Christians in his campaigns would be revealing. It would show, I think, great efforts at actually living life by principles, communication of views and beliefs, and especially disciplined simplicity. At the same time he was tolerant of differences. The concept of Truth became a way of finding common ground for those of differing cultures, religions and systems of belief. Would an explicit focus on Truth serve us as well? How might exploration be done between such very different cultures and languages?

"Ahimsa" – Nonviolence

The word "ahimsa" is literally and narrowly translated as "nonviolence" though I have also used the term "nonviolence" to refer to the totality. By "ahimsa" Gandhi refers to both the practice of noninjury to any living things, and to thoughts and intentions that lead to noninjurious acts. "Ahimsa" implies not only not initiating injury of another but also not responding to violence with violence. It includes not acting in a way that provokes violence. "Ahimsa" implies defenselessness. It also means voluntary poverty, for Gandhi believed that owning more than needed or eating more than needed is violence to those that do not have enough. Gandhi saw "ahimsa" and "sat" (truth) as integral and intertwined, but also that "ahimsa" is the means and Truth the end.

It seems to me that most of us at Wollaston Lake had some grasp of the principles of noninjury, though we did not inquire very far into it and would have uncovered many differences among us if we had. Meat eating vs. vegetarianism would have been an issue. Obviously none of us went as far as the Jains, whom Gandhi may have had in mind when he discussed the practice of noninjury, such as wearing face masks to avoid ingesting insects. What about insect repellent? Most of the Jain practices seem ludicrous to us. They come close to totally immobilizing a person from an active life. It must be said that Gandhi and his followers themselves failed to attain their own ideals, yet were criticized for even their modest efforts by the political pragmatists for going too far.

Our actual behavior was comprised of a stated willingness to not injure and to not respond violently so far as physical acts were concerned, but some made overt provocative gestures and used provocative words toward police, guards and Eldorado. Also, we practised destruction of property. One heard "some property has no right to exist." Getting to agreement on this one would take a very long time. Is it fundamental enough that we should take whatever time it requires?

Gandhi frequently called off events in which provocation was practised. It is useful to speculate about how our actions measure against his standards. I for one would like to know how Band members, who use firearms in hunting, understood the prohibition on their presence and use. Those of us for whom arms prohibition is a needless requirement, since we don't know how to use them and never think of them or possess them, have much to learn from those who do use them normally and choose not to.

For me, this brief discussion of "ahimsa" raises two main abstract questions. The first concerns asceticism (denial of material satisfaction). Gandhi was an ascetic in the Hindu tradition, though certainly not as a total vocation as is found among Hindu sects of several sorts. Asceticism is quite foreign to us, although there are plenty of precedents in Judaism and Christianity. I think it is quite possible that we would have been willing to undertake this sort of search for purity and perfection, if a reason for it could be established. But this is a big IF. Gandhi, however,

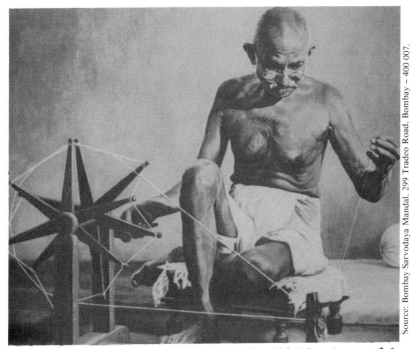

"My life is my message" -- *Mohandas Gandhi (above), one of the most important non-violent peace activists of the 1930's.*

was adamant about this and persisted in it. And some would claim that he succeeded because of it. Asceticism in modern life is foreign in a consumer society that thrives on indulgence, even narcissism. This modern rejection of it constitutes the basis for appeal of many modern cults that honor its absolute demands.

The second question that I have concerns the notion of obedience or authority. What a strange word obedience is to us! The leaders selected by Gandhi himself had great powers. They could ask participants to withdraw. They could negotiate independently of members. They could select or change goals. Their means were always patient reasoning and interpretation, and particularly, exemplification of the principles of nonviolence. Yet these leaders required obedience, particularly in the active phases of a conflict. I stress the word "obedience" and not "discipline" though the later, learned selfcontrol in a social role, was an integral part of the training of a "satyagrahi."

The Wollaston group was fairly disciplined, insofar as individuals gained some understanding of what was to be done and

271

Photo: Marie Tutt

acted in the pursuit of group goals. There were some departures from this discipline on the basis of imperfect understanding of what was to be done or imperfect agreement on tasks. Lack of clarity due to difficult communications was common. Any possible discipline easily breaks down under such conditions. There was also much impulsive behavior without reference to the input of others. So, in several dimensions there was lack of discipline. Of obedience there was none because there was no one to be obedient to, except the decisions of the Chief regarding major strategies, to which we had given deference in advance. We did not in fact discuss the issue of formal leadership, other than agreeing to some affinity group leadership that was imperfectly understood. Perhaps more significantly there was explicit rejection by some of authoritarian structures of all kinds. We have a lot to discuss here. Regarding obedience the gap between us and Gandhi is very wide indeed.

Chastity

Of all the principles that Gandhi insisted on he had the most trouble getting this one across. I wonder how a discussion of it would have gone at Wollaston Lake.

Nevertheless Gandhi insisted on chastity. His word was "brahmacharya." It is useful for us to consider his reasoning. He

Jack Ross at the mine gate on June 14, 1985, the first day of the blockade.

"I think that there was a lot of disagreement on what nonviolence means. I felt very threatened by my fellow blockaders who repeatedly engaged in copbaiting. There was a lot of hostility on our part, depending on the person and the situation. To me nonviolence should mean the expression of love for the opponent while opposing their values and positions. I can understand somewhat why this hostility existed, but I think that those that expressed it so openly should have been asked to utilize restraint and to turn their feelings into more constructive channels; or to withdraw. And there is a more personal issue than the feeling that such hostility leads to failure; the more the opposition is baited, the more likely I will get a lump on the skull."

– Jack Ross, June 24, 1985.

thought that a person who was wedded to pursuit of Truth could have nothing to do with gratification of the senses. Realization of Truth through self-gratification was a contradiction in terms. Marriage requires total devotion of husband and wife and excludes total devotion to universal love. Gandhi's own struggles with celibacy are well known. The meaning of chastity for nonviolence becomes clearer when we note that both Truth and "ahimsa," as defined above, call for complete purity of thought. They are inner matters. Control of sensual and lustful feelings then becomes the ultimate test. Gandhi extended "brahmacharya" to control of all sensual gratification, particularly eating. Fasting, for him was not just a weapon or a source of influence, but a means of "brahmacharya." From purity comes effectiveness.

Perhaps the notion of chastity is the one area in which modern conceptions of nonviolence have departed most from Gandhi. So far as religious ideals are concerned, sex is more often seen as the ultimate expression of love than as an evil to be suppressed. Was Gandhi wrong? How are we to explain some more recent successes with nonviolence that made no issue of chastity? Perhaps it should be noted that all great religions have experienced tensions over this issue and have provided uneasy solutions to it by such practices as celibate priesthood, phased abstinence and orgy, and numerous others.

We did not discuss or attempt to regulate sensual conduct at Wollaston. I have the feeling that it would have been fairly far down on our list of priorities if we had. But ultimately the issue has to come into the agendas of all longterm efforts.

Gandhi's boldness and candor about the subject of chastity was unprecedented in his era, and we must eventually come to terms with it. His success with nonviolence based upon it cannot be denied.

Nonpossession

This means living with faith that actual needs will be met. Gandhi believed in owning as little as possible. The person who is poor in goods cannot be threatened by robbery, confiscation or taxation. Gandhi sought total fulfillment of his wants only by wanting less. Furthermore, possession was seen as the temptation of thieves. The society in which thieving was impossible could be achieved by meeting all needs, sharing everything and eliminating temptation. The human body is also an object that

Looking south from the mine gate on June 14, 1985; on the left RCMP Officer Reynolds is standing beside his vehicle.

The RCMP regularly checked out the blockade area. The intimidating manner in which they handled themselves resulted in harsh words being spoken by both sides on a couple occasions. This happened despite the non-violent nature of the blockade being emphasized throughout preparations.

should be used for the pursuit of Truth and not as a possession. Gandhi also believed that nonpossession of thoughts should be practised. Education should be confined to useful things in the service of Truth.

Nonpossession is perhaps the least controversial and significant of Gandhi's principles, relevant more for a way of life than for a limited project. The basic principle, I think, is that property should not be allowed to stand in the way of our pursuit of Truth. I think that we understood this version at Wollaston Lake pretty well. The easy sharing of food and property and the willingness to go without comforts were taken in stride. This too was one of the things that succeeded best between the outsiders and the Band members. Still, property did sometimes become an issue. It was my automobile. I slept in it and did not like the way that the young men from Wollaston sat on my car's hood. We

275

outsiders have a lot to learn from the Indians about the nature of property. This too should be on a future agenda.

"Swadeshi" – Making And Using Local Goods

A note about "swadeshi" in closing. "Swadeshi," making and using local goods, was one of the bases of Gandhi's economic program. His stress on spinning was supported by his interest in the discipline of work, by the need for independence from British goods and by the need for a symbol of unity. He wanted to develop self-reliance to achieve independence from British goods so that British control could be challenged.

I would like to call attention to what every third world revolutionary economist has had to come to grips with: local production for direct consumption is a truly revolutionary force. Could outsiders to Indian reserves find some symbolic or substantive way to experiment with such an idea? I fear it would be easy to be the visiting dilettante on this. Have I underestimated what the Bands already do with this? Can Gandhi teach us anything here? What can the Native people teach us? If their dependence on government welfare systems was one weak point in their ability to sustain radical action against a government corporation, how can we contribute something to the trend to self-sufficiency? A big issue indeed. Perhaps it is not for us.

Concluding Remarks

This ends my discussion about Gandhi's relevance to the uranium blockade. It has been in part a scholarly exercise but also one of the imagination. I have frequently thought about our evaluation session after the blockade was lifted on June 17th. I have also tried to imagine each of the participants I knew and have tried to speak to you. I wish we could sit around the wood stove in my kitchen and explore it together. Thoughts flow well there. Let this essay be one part of the best substitutes I can devise for now.

CHAPTER 4
INTERNATIONAL SOLIDARITY

SAND
Scandinavians against nuclear Development

NUCLEAR DEVELOPMENT IN
SASKATCHEWAN + SCANDINAVIA

PUBLIC MEETING

Speakers, Performance + Refreshments

INDIAN METIS FRIENDSHIP CENTRE

168 WALL STREET SASKATOON

SUNDAY 7:30 P.M.
AUGUST 18 1985

Introduction

On August 1, 1985 a group of 11 people from Scandinavia arrived in Saskatoon. The group, who became known as Scandinavians Against Nuclear Development (SAND), included representatives from Norway, Sweden, and Finland. Amongst the group of nine women and two men were four teachers, three freelance journalists, a sailor, secretary, artist, and social worker. The main purpose of the visit was for the Scandinavians to see first hand the effects of uranium mining on the people and land. Their three week tour included the communities of Ile a la Crosse, Pinehouse, La Ronge, Southend and Wollaston Lake.

Saskatchewan uranium accounts for about half of Sweden's needs and about a quarter of Finland's. The 12 nuclear reactors in Sweden require about 1400 tonnes of uranium per year. According to the Swedish government, most of their Saskatchewan supply comes from Key Lake, although some uranium has been contracted for from Rabbit Lake and Collin's Bay. No contract documents have been made public.

Also in August 1985, Adele Ratt from La Ronge, Saskatchewan travelled to Japan to participate in numerous events commemorating the 40th anniversary of the bombings of Hiroshima and Nagasaki, including an international anti-nuclear conference. Her trip was sponsored by the Japan Congress Against A and H Bombs (Gensuikin).

In September 1985, Rune Eraker and Helge Hummelvoll from Norway arrived in Saskatchewan to gather information for a multi-media slide presentation. Their final production, called "Iktoms Profeti," was completed in February 1986.[1] It is a documentary about the conflict between uranium mining in northern Saskatchewan and the traditional Indian lifestyle. The English

279

Adele Ratt speaking at an international anti-nuclear conference in Japan August, 1985 (below), left to right: Nick Elmann, National Coordinator of KMU, the Philippine May 1st Trade Union; Manami Suzuki, Gensuikin; V.C. Mohan, Coordinator of Friends of the Earth Malaysia; Adele Ratt.

translation of the title is "Iktom's Prophesy." Iktom was a Sioux legendary figure who warned of what would happen when the white man came to North America. The show has toured Norway, Sweden, and Finland. It is 50 minutes long and contains 700 slides shown by 4 projectors.

280

iktoms profeti

– et multimedia-program om
atomindustriens bidrag til
ødeleggelse av livsgrunnlaget
til indianere i Nord Canada

Av Harald Eraker · Helge Hummelvoll · Rune Eraker

The front page of a leaflet made in Norway. It reads: "Iktoms Prophesy – a multi-media program about the nuclear industry's contribution to the destruction of the Indian way of life in northern Canada."

281

Scandinavians Against Nuclear Development (SAND) Tour: August 1985

A CBC TV reporter interviewing SAND members, from left to right, Raimo Long (from Finland), Karin Qvarnström (from Sweden), and Kia Lundqvist (from Finland) in Saskatoon, August 3, 1985. Moments latter the group departed for the north. The group visited Canada from August 1-19, 1985.

The group camped at Cole Bay just outside of Ile a la Crosse the nights of August 3 and 4, then stayed in Ile a la Crosse at peoples' homes the night of the 5th. On the morning of Hiroshima Day, August 6, Karin Qvarnström and Kia Lundqvist were interviewed on the Ile a la Crosse local radio.

The next stop on the tour was the 600 person Metis village of Pinehouse, the northern most community on the road to the Key Lake mine. Pinehouse Mayor George Smith and Birgitta Ohlsson from Sweden are deep into a discussion.

283

In the uranium boom-town of La Ronge, a small group gathered for a public meeting the evening of August 7th in the Indian and Metis Friendship Center.

Hiroshima and Nagasaki Commemoration In La Ronge

On August 8, 1985 SAND members and supporters held a vigil in front of the Key Lake Mining Corporation office in La Ronge to commemorate the 40th anniversary of the Hiroshima and Nagasaki bombings, respectively August 6th and 9th, 1945.

CBC radio in La Ronge interviewed a few members of the SAND group and Allan Quandt, a La Ronge resident, during the vigil. Below is a transcript of the radio program aired the same day.

CBC: La Ronge has visitors from far away in town today but they're not here for the fishing or the scenery. Their purpose is more serious than that. They're a group of observers and journalists from Sweden and Finland and they're concerned about the uranium mining here in northern Saskatchewan. I spoke with several of them in front of the Key Lake mining office this morning where they're holding a peaceful demonstration. What's your name?

Patrik Düring (PD).

CBC: And you are from?

PD: From Germany but I'm living in Sweden.

CBC: What brings you to La Ronge, Saskatchewan, Canada?

PD: Because in Europe, Sweden and Germany, we have nuclear power stations, and in Germany we have the Pershing rockets, and now it's 40 years ago that the Hiroshima and Nagasaki was, and Germany also is a part owner of the Key Lake mine.

CBC: Why are you opposed to uranium mining?

PD: Because we don't need uranium for energy and it's too dangerous. The risks are too big.

CBC: Why is it dangerous?

PD: Because the danger from radioactivity in the waste lasts millions of years, so our children and people who come after us have all the waste and we don't want that.

CBC: What is your name?

My name is Maj-Britt Andersson (M-BA) from Sweden.

CBC: What brings you here?

M-BA: Many things. We have many problems at home where I live with the waste from our nuclear power stations. In Sweden we have 12 nuclear power stations. We have many problems building big storage areas for the waste under the ocean and deep in the ground. That is what we are struggling against at home where I live.

CBC: What is the connection with Saskatchewan in that respect?

M-BA: The uranium is from here. Sweden buys over 52% of its uranium from Saskatchewan.

CBC: And why are you opposed to uranium mining?

285

On August 8, 1985 a vigil was held in front of the La Ronge Key Lake Mining Corporation office from 8 in the morning to noon, - the company's business hours.

M-BA: Because we don't want the nuclear power stations and all the waste. And because of the connection between the uranium and bombs, it's the same source. And then about the risks. We know that you in this country have the biggest risks to take because there is radium and thorium forever in the future in this country.

CBC: Who am I talking to now?

You're talking with Ritva Kovalainen (RK) and I come from Finland. I am a photographer and I also do some work for Finnish radio. I am here because Finland buys uranium from Saskatchewan.

CBC: Do you have any problems with that?

RK: Yes, we have problems with the waste. We don't know what to do with the waste we get from the power plants. In Finland we

286

On the far right a reporter from the La Ronge radio station interviews Karen Qvarnström. La Ronge resident Allan Quandt is standing behind her.

have four power plants. Two of them we bought from Russia and two from Sweden. From the ones we have bought from Russia we send the waste to Russia but the ones from Sweden we don't know what to do with the waste. And also we have heard in Finland that there might be some problems here with this uranium mining and we want to know what really happens. Now when we have been going around and interviewing all kinds of people, such as: politicians, writers, workers, unemployed people. We have really heard that it isn't so safe as our government has told to us.

CBC: The European visitors to northern Saskatchewan have been joined by several people from other communities in the north and from other parts of the province. One of them is a man well known for his views on peace and uranium development, Allen Quant. I asked him what he thought of this mornings events.

Allan Quandt (AQ): I think that it's marvelous that these people

287

have come a long way to see just exactly what the conditions are here. I think that their concern is the fact that there's nuclear development in their country, not related to any military installations but basically for energy. Because they don't belong to any military pacts like NATO and NORAD, but they're concerned because of the environmental effects and the effects upon people. And also they do feel strongly about peace and it's particularly significant because it's the 40th anniversary of the dropping of the bomb on Hiroshima. And so I think it's marvelous that they are making this attempt to link up with other groups and pay this visit to Canada.

CBC: Do they have just reason for all this anxiety?

AQ: Well, I think that they do have just reason. They tell us that their companies get their uranium from Key Lake. So there's a link. And they've asked to go to Key Lake and apparently have been denied to go and see the installation. I can't understand this. If the Key Lake people feel they haven't anything to hide and run a good operation, I can't understand why they don't let them in there.

CBC: What's your own feeling? Should uranium mining be stopped because of these concerns?

AQ: Absolutely.

Swedish sticker reading, "No To Uranium Mining."

288

From La Ronge the SAND group travelled north to Southend for a two night stay.

289

On August 10, when the Scandinavians arrived at the Eldorado mine gate, six men in four vehicles were waiting. The men would not say a word to the group and stayed close to their vehicles. Below, Stephanie Sydiaha from Saskatoon (left) and Maj-Britt Andersson from Öregrund, Sweden (right).

In Wollaston

On August 12, 1985 three SAND members made the following short statements on the Wollaston local FM radio station.

Maj-Britt Andersson (Sweden):

From the 12 nuclear power stations in Sweden comes very much waste every year. Where I live they are now building a big storage area for the waste. It is 5 meters under the ocean, then 50 meters under the ground. There are three big nuclear power stations not far from my home. I can see them every day. Below the sea in front of them the big storage area is being built.

At first they built two big tunnels that are 1,500 meters long each. And then two big tunnels 200 meters long each, then seven tunnels 100 meters long each, and four very big silos. One silo is 60 meters high and 30 meters wide. In the future, thousands of truck loads, each almost as big as your barge here, will drive down in the tunnels and put the waste in those caves under the sea. After perhaps 40 years they will shut the storage and start to pump out the ground water. Then it will be filled up with water and then all the dangerous materials in the waste will come out in our sea where we want to fish and swim. Those materials are dangerous to people for many thousands of years into the future. That's one reason why we struggle so hard against uranium and all energy you can make from it. They also can make nuclear bombs from the waste and these bombs can destroy the whole world.

I am a teacher and from our classroom window my pupils and I will see the big boat that comes with all the waste to this storage facility. In case there is a big accident from a nuclear power station or with the boat carrying the waste the government has already given us pills to take. But that cannot help us. And we think it is very ugly to think of, and to know, that we must have these pills in our schools and even in our homes.

These are some of the reasons why I came here. Our problem starts here. Uranium is the source for nuclear bombs and for uranium energy in many countries all over the world.

Birgitta Ohlsson (Sweden):

I have been struggling for many years against the nuclear industry in our national organization in Sweden called The Peoples' Campaign Against Nuclear Power and Nuclear Weapons. That struggle includes of course also trying to stop uranium mining. I believe that nuclear development is the most evil, dangerous and threatening thing we have in the world right now, and it all comes

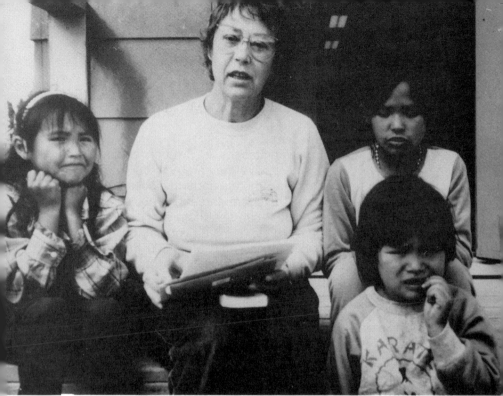

Birgitta Ohlsson and Wollaston children on the steps of the Local Advisory Council (LAC) office.

from uranium.

In my home area there are 13,500 inhabitants. We have especially been struggling against a planned nuclear waste storage site, the most dangerous waste that comes from the nuclear power plants, the spent fuel. It is dangerous for thousands of years. Among other things it contains plutonium which is a very poisonous material. The waste is so dangerous you can't even touch it. All over the world they don't know what to do with this waste. The have to keep it separated from all living things forever.

In Sweden the nuclear companies want to put the waste 500 meters down into the bedrock in big storage rooms. One day in 1980 the company came sneaking into my home area to start drilling in the bedrock to find a place to put that dangerous material. Nearly all the people in the area were against it when they found out what they were doing. First we tried to talk with the company and all the politicians but they didn't listen to us.

So after the drilling had been going on for several weeks we blocked the road to stop the water transport that they needed for the drilling. The politicians were still silent. About 200 people took turns blockading, 30 or 40 at a time. There were all sorts of people: men, women, youth, farmers, forest workers, a lot of teachers, doctors, dentists, and so on. After three days the police came. The company had called them, but we didn't move. At that time there was about 30 of us, including my husband and I. The policemen threw us into the snow drifts, but carefully because some of them agreed with us.

Then after a week another teacher and I had to go to court. We were found guilty and appealed to the highest court in Sweden. In total 25 people had to go to court. We were all found guilty and had to pay fines totaling 33,600 Swedish crowns, about $6,000. Many people from all over Sweden sent us money and we got more than enough to pay the fines. After that the company told us they wouldn't put the waste in that place. But they are still drilling in other places. In every place people are protesting.

I believe that uranium mining and the whole nuclear industry is a crime against humanity. When the men of power continue to destroy our world we have to protest, in many ways. And one of these ways is civil disobedience, that is without violence. The violence comes from uranium mining.

Vivianne Johansson (Sweden):

Also in my home area we have struggled against planned waste storage and we have succeeded. The government wanted to drill down into the rock and make an underground room where they could put the spent fuel from the nuclear reactors. We started to guard that area every day and night. We lived in two small trailers in different places. At least two people stayed in each trailer. All together there were a lot of people who took turns guarding. After a while we even got telephone, electric lights and our own post address to the trailers. We started in 1979 and have now guarded that area for five years. Every day four people have been there guarding and have now succeeded. The authorities are not going to put the waste there.

I am very glad to be here and meet the people of Wollaston and to see how beautiful it is where you live. Thank you very much, from the Scandinavian group.

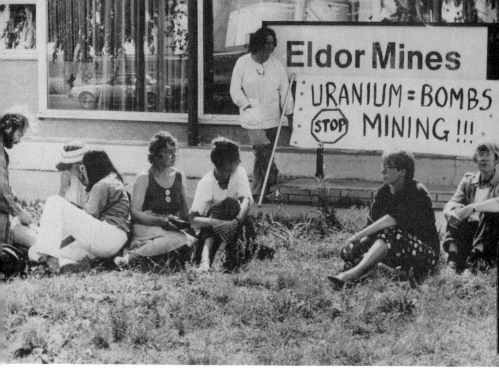

At the Eldorado Nuclear Ltd. office in Saskatoon, August 16, 1985.

Back In Saskatoon

On August 16, 1985 SAND members visited the Eldorado Nuclear Ltd. office in Saskatoon to request Eldorado's participation in a public meeting. Eldorado refused and asked the group to leave the office immediately. The SAND group complied, though Nicole Laurendeau and Miles Goldstick stayed to try and speak to the management. The receptionist directed them to, "Please have a seat and wait." Soon the police arrived, arrested the two and laid charges of "mischief." The Saskatoon CBC television news called the event "a sit in," and did not report Eldorado's consistent refusals to meet with any opposition in a public forum. At Eldorado's request, the charges were dropped a few days before the scheduled court appearance in February 1986.

Despite Eldorado's refusal to participate in a public meeting with SAND, a public meeting was held the evening of August 18, 1985 in the Saskatoon Indian and Metis Friendship Center. The poster on page 278 was made to advertise the meeting. About 50 people attended.

294

Swedish version of the international anti-nuclear symbol.

SAND At Home

Once the SAND group returned to their homes they spread the information they had gathered on the tour. A number of newspaper, magazine and book articles were published in Sweden and Finland.[2] Ritva Kovalainen and Kia Lundqvist had a full page article published on October 13, 1985 in *Hufvudstadsbladet*, one of Finland's largest Swedish language newspapers. They also had three radio programs aired on Finnish language radio and one on Swedish language radio in Finland. Karin Qvarnström from Stockholm produced a slide show that has been used by public interest groups in Sweden and Finland.

The following letter was distributed in August 1985 to the mass media in Sweden by the Swedish members of SAND. The letter was written in Wollaston Lake on August 15, 1985.

At the right is a reduction of a full page newspaper article published in the Swedish language Finnnish daily Hufvudstadsbladet. *The headline reads,* Canada: The local people near the mines are protesting; the buyer also has responsibility for the uranium. . . *The text under the centre photo on the left is,* "The mining company only cares about money. For them, people are less important," says Martin Smith. *Under the photo on the right is written,* "The uranium mines don't do anything good for us Indians in northern Canada," says George Smith who is the Mayor in Pinhouse.

296

Kanada: Lokalbefolkningen vid gruvorna protesterar

Också köparen har ansvar för uranet ...

Urangruvan för ingenting gott med sig för oss indianer i norra Kanada, säger George Smith som är borgmästare i Pinehouse.

on, nor-
tchewan,
Tre vägar
rut. Vid
varje väg
urangru-
Lake,
e, Rabbit
nyöpp-
in's Bay.
un slutar
en som
e norrut
yget.

an är en
en av Kana-
breder den
prärien ut
norra de-
insen är be-
mager skog
a sjöar. Alla
na till var-
digaste sjö
norrut via
ver till Isha-
e sydligare
ut i Hud-

dianer

i provinsen är
Av dem bor
norra Saskat-
teten av be-
består av cree-
mestiser
har vuxit fram
av blandkulten-
och indianer
n räknas som
tig.

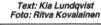

Indianerna har många barn. Av byns 600 invånare är sextio procent under tjugo år.

Största urangruvan

☐ Den kanadensiska del-
staten Saskatchewan pro-
ducerar över 15 procent av
allt uran i världen. Detta
gör Kanada till världens
främsta uranexportör. Det
kanadensiska uranet säljs i
första hand till USA men
en stor del går också till
Europa. Också Industrins
Kraft köper uranet till
kärnkraftverken i Olkiluoto
från Key Lake-gruvan i
norra Saskatchewan.

Key Lake-gruvan anses
vara den rikaste uranfyndig-
heten i världen. Gruvan
har rykte om sig att vara

modernast och säkrast i
världen.

Trots det har gruvan haft
åtminstone 14 kända öns-
kade utsläpp under åren
1983–1985.

I januari 1984 över-
svämmades avfallsreservoa-
ren vid gruvan och 100 miljoner liter radioaktivt avfallsvatten rann ut i om-
givningen.

☐ En svensk-finländsk
grupp reste omkring i norra
Saskatchewan under tre
veckors tid i augusti. Dele-
gationen bestod av elva
personer, som var och en
hade bekostat resan själv.

Deltagarna träffade in-
dianer, mestiser och vita
som alla var oroade av
uranindustrins intrång i området.

Ändamålet med besöket
var att försöka skaffa allsidig information om kärn-
kraftcykelns första skede,
uranbrytningen.

Ansatsen till insamlandet
av allsidig information
strandade på kanadens-
tribolagens totala ovilja att
lämna ut någonsomhelst information.

Gruppen vägrades bl.a.
besöka Key Lake-gruvan.

angruva måste man komma upp
ur brottet med vissa mellanrum
för att inte få för mycket strålning. I Key Lake-gruvan får ar-
betarna göra överlappa skift
utan att någon säger till dem att
komma upp.

Vid Key Lake är man enligt
min uppfattning oansvarigt i
förhållandet till gruvarbetarna.

George Smith berättar att
gruvarbetarna i kåkel till blir på
sig mätare som visar hur mycket
strålning den enskilda arbetaren
utsatts för.

De kan inte själva avläsa mätarna. Bolaget skickar mätarna
någonstans för att avläsas, men
meddelar aldrig gruvarbetarna
resultatet. Man kan lita på att
allt går rätt till?

Regeringsägda

Gruvbolagen som bryter
uran i Saskatchewan äga alla
till största delen av de federala
och provinsiella regeringarna.

Gruvbolagen fungerar enligt
lokalbefolkningen mycket egen-
mäktigt. År 1979 blev det Lex.
känt att Key Lake Mining Corporation hade torkat ut 16 sjöar
för att anlägga urangruvan vid
KeyLaten. Bolaget hade inte in-
förskaffat tillräckliga lov och
tänndes därför inför rätta. Bolaget fick böta 500 kanadensiska
dollar, ca 2 500 mk.

– Sedan dess har vi haft fjor-
ton olika utsläpp vid gruvan, berättar Peter Prebble som är
medlem av The Interchurch
Uranium Committee (IUC),
Samkyrkliga urankommittén i
Saskatchewan.

Text: Kia Lundqvist
Foto: Ritva Kovalainen

radioaktiv strålning är far-
lig.

Strålningen går varken att se
eller känna och dessutom varier
den över en lång tid. Skadorna
syns kanske först om 20–30 år.
Allt detta vålar naturligtvis
oro hos lokalbefolkningen.

Invånarna i indian- och mes-
tissamhällena i norra Saskatchewan talar antingen cree eller
dene som modersmål.

I dessa språk finns inget ord
för strålning. Utbildningssitua-
tionen är också skral. Någon
utbildning utöver det allra ele-
mentäraste står inte till buds i
samhällena. Många personer är
nätt och jämnt läs- och skriv-
kunniga, medan en del är anal-
fabeter.

Men de lyckas alltid vända sina ord så att det ser ut som om

I kommittén finns alla kyrkor
av betydelse i Saskatchewan re-
presenterade. Dessa är bl.a. de
lutheranska, katolska, angli-
kanska och mennonitiska kyrkor-
na samt en specifikt kanadensisk
kyrka; United Churches.

Kommittén motsätter sig
kärnkraftsindustrin i sin helhet,
men koncentrerar sig främst på
att motarbeta uranbrytningen i
Saskatchewan. Kommittén vill
informera allmänheten om
uranbrytningens häkscrisker och
om hur uranet slutligen ofta an-
vändes som råmaterial till kärnvapen.

Redan under de fyra
första månaderna hade
Key Lake-gruvan många
utsläpp. I januari 1984
rann c. 100 miljoner liter
radioaktivt avfallsvatten ut
i omgivningen.

Gruvan ligger långt ute i öde-
marken men det finns ett litet
uttal indianer som jagar eller
som på annat sätt får sin ut-
komst ur området i närheten
av gruvan, säger Peter Prebble.

Bolaget hade tur då det stora
utsläppet skedde. Det var vinter
och vattnet frös snabbt.

Hade det varit sommar hade
olyckan fått helt andra dimen-
sioner. Bolaget hade också tur i
ett annat avseende. Det av-
renade vattnet runnit ut i en
större sjö som är i direkt kon-
takt med hela vattensystemet i
norra Saskatchewan, fortsätter
Prebble.

Sjöarna i Norra Saskat-
chewan är alla anslutna till
varandra. Problemet är att
om radioaktivt avfallsvatt-
ten från urangruvorna
kommer ut i vattensyste-
met så kan ingen förutsäga
hur stora områden har
orenas. Om vattnet för-
orenas blir indian- och
mestisbyarnas utkomst-
möjligheter ytterligare
kringskurna. Fisket är
nämligen viktigaste ut-
komstkällorna i dessa byar.

Sker överallt

ICU ser enligt Peter Prebble
utsläppen vid gruvorna som ett
bevis för ett allmängiltigt problem i samband med uranbrytningen:

– Vi har velat understryka
att utsläpp som sker vid Key
Lake sker överallt i världen där
man bryter uran.

I New Mexico, USA har stor-
ra vattensystem förorenats p.g.a.
radioaktiva utsläpp från urangruvorna. Boskap har förgiftats
och människor har inte kunnat
dricka vattnet i området. Lik-
nande olyckor har också skett i
Australien.

Uranbrytningen utgör
det första skedet i kärn-
kraftscykeln. Från de ställ-
len där uranet bryts, trans-
porteras det vidare ut över
världen för att raffine-
ras, anrikas och användas i
kärnkraftverk eller för
kärnvapen. Slutligen måste
avfallet uppbevaras.

Inget land som matar sin in-
dustri och sina privata hushåll
med energi från kärnkraftverk,
kan avsäga sig ansvaret för de
olika skedena i kärnkraftscy-
keln.

Användningen av kärnkraft
är ett globalt problem, som i
detta fall börjar vid Key Lake i
Saskatchewan och slutar med
problemen kring avfallshante-
ringen vid Olkiluoto i Finland.

'Inte säkert'

– Det är min uppfattning
som indian, att förvaringen av
avfallet inte är säker.

Som jag ser det, bryr de sig
inte om människorna. De bryr
sig endast om pengarna de får
från uranbrytningen. För dem
är människoliv mindre viktiga,
säger Martin Smith från Pinehouse, som än dag fram till 1982
arbetade vid urangruvorna.

George Smith tillägger:
– Vi har alltid försörjt oss.

– Också när gruvorna stängs,
kommer vi att leva vidare och
försörja oss här i norr. Detta är
ju det bästa stället i världen.
Det här är ju Guds land.

dianer

till stora de-
lo, hjort,
nns i rikliga

ren är näs-
folkningen i
ewan har att

är få.
98 procent av
arbetslös. Både
alösheten och
befolkningen i
sultur som är
na pengar
har jakten,
g av fällor,
odlandet av
för betydelse

stil

orra Saskat-
nu att hela
ivsstil är ho-
de av federa-
regeringarna
med multina-
tidigare samt
de skogsbe-

ever det som
ela deras lev-
modern form

uranbrytning-
på skogen
onomisk och
ukturen
sätts för, har
utan något
inflytande.

– Gruvbolagen bryr sig endast om pengarna. För dem
är människoliv mindre viktiga, hävdar Martin Smith.

en sker ...

van innan man får produkten
"yellow cake" som transporte-
ras söderut för att anrikas till ur-
van. Man senare kan användas
antingen som bränsle till kärn-

öppna dagbrott. Själva uranet är endast en liten del
av uranmalmen.

Uranmalmen genomgår
många olika processer vid gru-

kraftverken eller för t.ex. kärnvapen.

Då uranmalmen förvandlas
till "yellow cake" får man som
biprodukt stora mängder både

fast och flytande radioaktivt
avfall.

Detta avfall förvaras i enor-
ma reservoarer vid gruvorna.
Det finns inga metoder för att

neutralisera eller göra avfallet
ofarligt. Avfallsområdena val
gruvorna är varken inhägnade
eller övertäckta.

Enligt vissa uppgifter är ocksä
brottens på reservoarerna outi-
fredsställande byggda.

Ett stort problem är att
många tungmetaller bl.a.
arsenik och kvicksilver fri-
görs från uranmalmen i
samband med gruvdriften.
Likaså kvar uranet endast ett
av de radioaktiva ämnen
som finns i uranmalmen.

Andra radioaktiva äm-
nen som lämnas kvar i av-
fallet är bl.a. radium och
torium.

Få nya jobb

På väg upp till Key Lake-gruvan ligger det lilla mestis-och
indiansamhället Pinehouse.
Pinehouse har ca 600 in-
vånare.

– Då Key Lake öppnades
lovade gruvbolaget att 50 procent av arbetsplatserna vid gruvan skulle besättas med människor från de närliggande samhällena. I dag jobbar endast 4
personer från Pinehouse vid
gruvan, säger George Smith
som är borgmästare i Pinehouse.

George Smith är själv indian
och livnär sig som de andra i
byn som påsljägare och fiskare.

Han fortsätter: Vi har försökt
förklara för myndigheterna att
urangruvorna inte för någonting
gott med sig för oss indianer. De
ger inga jobb och gagnar inte
heller vårt samhälle i andra av-
seenden.

vi skulle få arbetstillfällen via
gruvan. Avfallsområdena val
gruvorna är varken inhägnade
eller övertäckta.

Enligt vissa uppgifter är ocksä
– Inga jobb. De har all makt.
Pengarna försvinner söderut och
lite av pengarna tillfaller oss.

Lite information

Gruvbolagen ger varken
gruvarbetarna eller den lo-
kala befolkningen tillräck-
lig information om farorna
med uranbrytningen. Mot-
ståndarna till uranbrytningen berättar att det är svårt
att förklara på vilket sätt

tt indian- och mestissamhälle på vägen upp mot Key Lake-gruvan.

Open Letter To All Political Parties
About Uranium Mining In Canada

In order to form an opinion about uranium mining we, 8 people from Sweden and 3 from Finland, have for three weeks travelled 2,500 km between Saskatoon and the uranium mines in northern Saskatchewan, Canada. The whole trip has been paid for by ourselves. We rented a small bus, camped or slept on the floor in community centres, and made our own food.

We have continually had the opportunity to speak with people at the different places we have visited. Politicians at all levels have been interviewed, and we have explained to them why we have done this trip. The TV, radio, and newspapers reported on us daily.

Canadian uranium was initially mined to supply the United States nuclear weapons program. Since then uranium mining has developed into a big industry. Exploration has taken place in large areas of northern Canada. The Indian way of life has been brutally ignored. Their lives and health are seriously threatened.

Many Indian people can neither read nor write and therefore have difficulties in understanding the danger uranium mining presents to all living things for millions of years. Even people with "first class" education do not understand the dangers.

The Canadian government together with foreign companies cynically continue mining. Some of the companies are actively involved in weapons production. Canadian uranium is sold in many countries, including many nations with nuclear weapons.

Sweden buys more than 50% of its uranium from Canada. Many Swedish politicians may be unaware of what uranium mining has done in Canada. For example, 17 lakes have been drained at Key Lake, which is the biggest mine. This is surely the result of no independent investigation having been done.

After our trip in Canada's wilderness it is impossible for us to accept that Swedish people in our aware country participate in taking uranium out of the ground. Uranium mining spreads radioactivity that contaminates the Earth forever. The nuclear weapons powers get the raw material they need for their missiles from uranium mines. Sweden does not contribute to peace by importing uranium.

We demand that the import of uranium be stopped!
Birgitta Ohlsson, Maj-Britt Andersson, Vivianne Johansson, Donald Wallström, Karin Qvarnström, Margareta Åkerström, Christine Rognerud, and Patrik Düring.

298

Afterword:

Rationale And Proposal For A Northern Development Council

By Adele Ratt, May 1985

Rationale

Northern Saskatchewan is increasingly being controlled by southern interests, namely government and multi-national corporations. The revenue generated by resource exploitation (most of which is taking place on Indian land) is not benefiting northern people; and the rate of unemployment in northern communities is sometimes 90%. We are all fully aware of the problems this has created, such as: violence and crimes of violence, native people filling up Saskatchewan jails, dependency on government social programs such as welfare, family breakdown and resulting problems such as alcoholism, wife beating, and child abuse and neglect. The list is long.

There is a need to develop an organization that will ensure northerners' control over northern development.

Proposal

A Northern Development Council should be elected by a majority of all northern community residents and include representation from Local Community Authorities (LCA s), Local Advisory Councils (LAC s) and Band Councils.

This elected body will approve all major new developments proposed for northern Saskatchewan by government and corporations; with close consultation with all communities, Bands, native organizations and residents of northern Saskatchewan.

299

Consultation should include wherever possible the views of women's groups, youth groups, fishermen, trappers, and elders' councils from each community.

This council will act as a monitoring body for all management in the areas of forestry, fisheries, wildlife, land use policy and tourism. It should have a policy of promoting self-sufficiency for all northern communities, Bands and residents of northern Saskatchewan.

It should work towards these long term goals:

1) economic independence and self-sufficiency of all northern communities and Bands,
2) land claims settlement for metis and non-status Indians and Bands,
3) protection of hunting, fishing and trapping rights of treaty and non-status Indian and Metis people,
4) guarantee of aboriginal rights,
5) meaningful and environmentally sound employment,
6) guaranteed water quality and environmental protection for northern Saskatchewan,
7) northern control of education and social systems,
8) restoration of cultural heritage,
9) social, economic, cultural and environmental security for future generations.

This council should set up and conduct all public inquiries into any major new developments being proposed for northern Saskatchewan.

This council should actively seek alternatives to:

– welfare and social programs,
– education and health programs,
– resource exploitation,
– law enforcement and law making,
– skills training and employment, and
– environmentally destructive technology.

To ensure a healthy, happy, secure and independent future we must all work toward educating people and getting people to be more socially aware of the issues we are facing in the north. A massive education campaign has to be conducted with the view to regaining control over our lives and promoting alternative lifestyles to the present, fast paced, ecologically dangerous lifestyle.

Northern people want to control their lives. Since the north has been colonized, assimilated, technologized, propagandized and forced to accept a way of life and thinking that is destructive and suicidal; it is going to take a massive effort and commitment

300

to achieve the goal of survival and independence.

It is never too late to look deep into ourselves and our lives, and ask, "What can I do to change the way things are?" We must accept our responsibility to our children and grandchildren. The decisions we make or don't make today will affect them for generations to come. It is up to us to ensure that they have a healthy, safe, secure future. It is time to work towards this goal.

Mother Earth

My eyes tell the story that has looked upon the
SPIRIT OF NATURE.
I have seen the mutilation put on Mother Earth.
I have shared her sorrow, I have shared her pain.
I am her flesh, we are as one.
For me to be without her, I'm lost.
She is the SPIRIT of my soul being.
When she struggles from all those wounds
that have been cut deep into her flesh,
I SUFFER....

—Adele Ratt, summer, 1985.

301

Footnotes

THE PEOPLE

[1] Dobbin, Murray (ed.). November, 1984. "Economic Options for Northern Saskatchewan, Report of the Economic Options for Northern Saskatchewan Conference, Saskatoon, November 16–17, 1984." 48 pages. See page 11. Inter-Church Uranium Committee and Northern Native Rights Committee. Available from: One Sky, 134 Ave. F South, Saskatoon, Saskatchewan, Canada. S7M 1S8. Tel. 306-652-1571. $4.50CDN.

[2] Environment Canada. February, 1981. "1980 Wastewater Compliance Report For Federal Establishments In Saskatchewan." 36 pages. See page 2. Western and Northern Region Environmental Protection Service, Environment Canada, Saskatchewan District Office, Regina, Saskatchewan.

[3] Ralph, Diana Ph.D. February 11, 1984. "Faulty Prescription For Northern Native People: Health-damaging Development And Little Care." 32 pages. In most cases, data in the report does not distinguish between northern native and non-native people. Thus, the life expectancy and death rate statistics mentioned have been extrapolated.

[4] Bouvier, Vye. February, 1985. "Cancellation of Government Food Subsidy Upsets Wollaston Residents." *New Breed*. Vol. 16. No. 2. Pages 8, 9 and 27.

[5] One Sky. January, 1982 (revised June, 1984). "Information Kit: Native People." 224 pages. See page 39. One Sky, 134 Ave. F South, Saskatoon, Saskatchewan, Canada. S7M 1S8. Tel. 306-652-1571. ISBN No. 0-9691449-03.

[6] Reynolds, Margaret. "The Dene Language Book." Available from the Saskatchewan Indian Cultural College, Saskatoon. 42 pages. $3CDN.

[7] His Most Gracious Majesty the King of Great Britain and Ireland. 1907. "Treaty Number 10 and Reports of Commissioners." Queen's Printer, Ottawa, Canada (reprinted in 1966). Cat. No.:Ci 72-1066. IAND Publication No. QS-2048-000-EE-A-11. See pages 10-13 for the Treaty.

[8] Government of Canada. July, 1970. "Lac La Hache Reserve No. 220. "Prince Albert Land Titles Office No. 67 PA 018251.

[9] The statements were made at a community meeting in Wollaston Lake on April 30, 1985 and in an interview made in Pinehouse on August 7, 1985 by SAND.

[10] This statement was made during an interview made in Ile a la Crosse on August 5, 1985 by SAND.

THE MINES

[1] Reguly, Eric. October 4, 1986. "Uranium Industry Bracing For Wave Of Protectionism." *The Financial Post*. Page 38.

[2] Salaff, Stephen, March, 1985. "The Cigar Lake Mine, the Real Drilling Begins At Saskatchewan's Prize Uranium Deposit." *Saskatchewan Business*. Pages 47-48. See page 48. The article does not give the name of the president nor his company.

[3] Moelaert, John. April, 1979. "This Dust Is Making Me Sick." *The Energy File*. 3 pages. Reprinted May, 1980 by: Regina Group For A Non-nuclear Society, 2138 McIntyre St., Regina, Saskatchewan.

[4] Eldorado Nuclear Ltd. March, 1984. "Annual Report 1983, As Submitted To The AECB." Page 41.

[5] Eldorado's 1984 annual report states the mill produced over 1.6 million kilos of yellowcake. The selling price would have to be $62.50CDN/kg for this amount to be worth $100CDN million. The actual price, which is kept secret, could be higher or lower.

[6] January 24, 1984. "Men Say Fired For Unionizing." *Star Phoenix*, Saskatoon, Saskatchewan.

[7] Pugh, Terry. September, 1986. "Garbage Never Looked So Good." *Briarpatch*. Page 5.

[8] Pugh, Terry. September, 1986. Ibid.

[9] Canadian Press. August 16, 1986. "AMOK Solves Problem of Radioactive Waste." *Globe and Mail*.

[10] Jorgensen, Bud. May 4, 1986. "Easing of Uranium Export Rules Urged for Canada." *Globe and Mail*. Page B8.

[11] June 19, 1985. *The Northerner*, La Ronge, Saskatchewan, Canada. S0J 1L0. Page 10.

[12] Menely, W.A. April 15, 1980. "Environmental Management in the Uranium Industry, Statement of Evidence to the B.C. RCUM, Phase VI: Environmental Impact." 11 pages. See page 8. W.A. Menely Consultants Ltd., Consulting Hydrologists and Geologic Engineers, Box 7167, Saskatoon, Saskatchewan, Canada. S7K 4J1.

[13] Information on the glories of the Key Lake mine is eagerly distributed free of charge by KLMC, 229-4th Avenue South, Saskatoon, Saskatchewan, Canada. S7K 4K3. Tel. 306-665-7000.

[14] Detailed accounts of the spill and clean up are available

from KLMC. See the note above for addresses.

[15] Op. Cit. Saliff, Stephen. March, 1985. Pages 47–48.

[16] Op. Cit. Reguly, Eric. October 4, 1986.

[17] Morland, Howard. 1981. "The Secret That Exploded." Random House, N.Y., N.Y., U.S.A. 289 pages. See pages 277-279. ISBN: 0-394-51297-9.

[18] Eldorado Nuclear Limited. May, 1985. "Information: Eldor Mines, Eldorado Nuclear Limited." 3 pages. See page 3. For more information write to: Eldorado Nuclear Limited, 400-225 Albert Street, Ottawa, Ontario, Canada. K1P 6A9. Tel. 613-238-5222.

[19] Most of this section is reprinted from: Miles Goldstick. August, 1984. "Uranium Bullets, There Is No Such Thing As 'Peaceful Uranium Mining'." 4 page Uranium Traffic Special Report. Available from Earth Embassy, Box 3183, Vancouver, B.C., Canada. V6B 3X6.

[20] Bolté, Philip, L. May, 1983. "The Tank Killers - Tungsten Vs. Depleted Uranium." 6 pages. *National Defense*, May/June, 1983. Seideman, Paul. Jan., 1984. "DU: Material With A Future." 4 pages. *National Defense*. Published monthly. American Defense Preparedness Association, Suite 900, 1700 N. More St., Arlington, VA, USA. 22209. Tel. 703-522-1820.

[21] Letter from Major General Peter G. Burbules (Department of the Army, Headquarters, US Army Armament, Munitions and Chemical Command, Rock Island, Illinois, USA. 61299-6000.) to William J. Dircks (Office of the Executive Director of Operations, US Nuclear Regulatory Commission, Washington, DC. 20555.) Undated, though stamped "Received, KMNC Marketing, 14 MAR 1985." Available from: Native Americans For A Clean Environment, Route 2, Box 51-B, Vian, Oklahoma, USA. 74962. Tel. 918-773-8184.

[22] Copenhagen, *Politikin*, Sept. 19, 1984.

[23] The information on Aerojet is from four pieces of their promotional literature, all undated: Untitled, color, glossy booklet. 16 pages; "Family of 30 mm GAU8A Ammunition." 2 pages; "Depleted Uranium Penetrators." 29 pages; and "Heavy Metals Division." 12 pages. Aerojet Ordnance Company, 2521 Michelle Drive, Tustin, California, USA. 92680. Tel. 714-730-6004.

[24] Lumsden, Dr. Malvern. 1975. "Incendiary Weapons." 255 pages; and 1978. "Anti-personnel Weapons." 299 pages. Stockholm International Peace Research Institute (SIPRI), Bergshamra, 171 73 Solna, Sweden. Tel. 08-599 700.

[25] Prokosh, Eric. 1972. "The Simple Art Of Murder, Anti-personnel Weapons And Their Developers." 88 pages. National

Action/Research on the Military Industrial Complex (NARMIC) - A project of American Friends Service Committee, 112 South 16 St., Philadelphia, Pennsylvania, USA. 19102.

[26] Ezell, Edward Clinton. 1977. "Small Arms Of The World." 11th revised edition. 671 pages. Stockpole Books, Cameron and Kelker Streets, Box 1831, Harrisburg, PA, USA. 17105.

[27] *Jane's Weapons Systems; Jane's Fighting Ships; Jane's All The World's Aircraft.* Published annually. Jane's Publishing Co. Ltd., 238 City Road, London, England. EC1V 2PU. Tel. 01-251-9281.

[28] Pohl, Robert O. August, 1976. "Health Effects of Radon-222 from Uranium Mining." *Search*, Vol.7, No. 8, pages 345-350. See pages 346 and 350. Robert O. Pohl, Cornell University, Laboratory of Atomic and Solid State Physics, Clark Hall, Ithaca, N.Y. 14853 U.S.A. Telex: 937478.

[29] Uranium-235 and thorium-232 (which is always found together with uranium) both change 11 times before becoming non-radioactive lead.

[30] For an explanation of radioactivity and its different forms see: Bertell, Rosalie, Ph,D., G.N.S.H. 1985. "No Immediate Danger - Prognosis for a Radioactive Earth." The Women's Press Ltd., 124 Shoreditch High St., London, England E1. 435 pages. ISBN: 0-7043-3934-X PBK.

[31] Ibid. Page 24.

[32] The assumptions made in this calculation are: mill rate = 600,000 tonnes ore per year; average grade = .3%; U_3O_8 is 85% uranium; 3.0×10^{-7} grams of radium-226 per gram uranium; $[600,000 \times .003 \times .85 \times (3.0 \times 10^{-7})] = 459$ grams of radium-226 produced by the Rabbit Lake mill per year, or roughly 1.25 grams per day of operation.

[33] Dreeson, D.R. Feb., 1978. "Uranium Mill Tailings - Environmental Implications." LASL Mini-review. 4 pages. See page 4. LASL-77-37. Los Alamos Scientific Laboratory, University of California, Los Alamos, New Mexico, USA 87545.

[34] Garrick, David. November 26, 1986. "Presentation to the NDP Inquiry: What Ordinary Canadians Think of Nuclear Energy." 8 pages. See pages 2 and 3. Earth Embassy, Box 3183, Vancouver, B.C., Canada. V6B 3X6.

[35] AECB, Advisory Panel on Tailings. Sept., 1978. "The Management of Uranium Mill Tailings - An Appraisal of Current Practices." 31 pages. See page 14. AECB Report 1196. AECB reports are available free of charge from: AECB, Box 1046, Ottawa, Ontario, Canada. K1P 5S9. Tel. 613-995-6941.

[36] Eldorado Nuclear Ltd. March, 1984. "Annual Report 1984,

As Submitted To The AECB." Page 36. Eldorado Nuclear Ltd., 400-225 Albert St., Ottawa, Ontario, Canada. K1P 6A9. Tel. 613-238-5222. Telex: 053-3382.

[37] A detailed list of chemicals used in both the acid and alkaline uranium milling process is in: U.S. E.P.A. June, 1974. "State of the Art: Uranium Mining, Milling, and Refining Industry." 113 pages. See page 54. Project No. 21AGF-02. Program Element 1BB040. EPA-660/2-74-038. U.S. E.P.A. Env. Research Lab., Box 1198, Ada, Oklahoma, U.S.A. 74820.

[38] Eldorado Nuclear Ltd. March 7, 1985. "Annual Report 1984, As Submitted To The AECB." Page 36.

[39] Ibid. Page 53.

[40] Contaminated water from the pit was pumped into Pow Bay via the drainage ditches from the beginning of mine operation in June, 1975 until October, 1977. In November, 1977, the mine water began to be used as process water in the mill.

[41] Leaflet distributed by the Inter-Church Uranium Committee, Box 7724, Saskatoon, Saskatchewan, Canada. S7K 4R4. Tel. 306-934-3030.

[42] This calculation is based on the conservative approximation that one cubic metre of wastes weigh one and a half metric tonnes, and that a two lane highway is ten metres wide. A ten metre wide highway 800 km long would be covered about 34 cm deep by 4 million tonnes of waste.

[43] The same conservative measures are used as in the note above. The distance from coast to coast is taken as 7,200 km.

[44] Swanson, Stella M. September, 1982. "Bioaccumulation of Radionuclides with Reference to Coal-fired Power Plants." 28 pages. See pages 1-10. SRC Publication No. C-805-48-E-82. S.M. Swanson, Saskatchewan Research Council, 30 Campus Drive, Saskatoon, Saskatchewan, Canada, S7N 0X1.

[45] Swanson, Stella M. Early 1982. "Levels of Ra226, Pb210, and Utotal in fish Near a Saskatchewan Uranium Mine and Mill." Health Physics. Vol. 45, No. 1 (July), pages 67-80, 1983. See page 67. S.M. Swanson, Saskatchewan Research Council, 30 Campus Drive, Saskatoon, Saskatchewan, Canada, S7N 0X1.

[46] Sheard, J.W. November, 1983. "Interim Report on Radionuclide Distribution in the Vegetation of Northern Saskatchewan to Saskatchewan Health Research Board." 26 pages. Department of Biology, University of Saskatchewan. A joint project with the Saskatchewan Research Council.

[47] Shook, E. and Sedgwick, W.H. April 14 , 1980. "Uranium Brief - Developed By The 'Uranium Committee' Of The Upper North Thompson Livestock Association from October, 1977 to

February, 1980." 16 pages. See pages 2-3. Statement of Evidence to the B.C. RCUM Phase VI: Environmental Impact.

[48] Dobyns, Douglas E. March, 1980. "Notes Submitted By Douglas E. Dobyns, Islands Protection Society, To The B.C. Royal Commission Of Inquiry Into Uranium Mining, Phase VI: Environmental Impact." 8 pages. In addition, a more detailed work was completed in 1985.

[49] Il'enko, A.I. 1971. "Radioecology of Wild Animals." In Radioecology, Klechkovskii, V.M.; Polikarpov, G.G. and Aleksakhin, R.M. (eds.). Russian translation by John Wiley and Sons.

[50] Ruggles, R.G. and Rowley, W.J. Nov. 1978. "A Study of Water Pollution in the Vicinity of the Eldorado Nuclear Ltd. Beaverlodge Operation 1976 and 1977." 82 pages. See pages 52-54. Fisheries and Environment Canada, Environmental Protection Service, 8th floor, 9942-108 Street, Edmonton, Alberta, Canada. T5K 2J5 Surveillance Report 5-NW-78-10.

[51] Op. Cit. Swanson, Stella M. Early 1982. See page 67.

[52] A chironomid is a midge, or small biting insect.

[53] Op. Cit. Swanson, Stella M. Early 1982. See page 78.

[54] Saskatchewan Research Council, Chemistry and Biology Division. December 13, 1978. "Environmental Overview Assessment for the Dubyna 31 - Zone Uranium Production Program, For Eldorado Nuclear Ltd." 140 pages. For water values see page 27; plants, page 44; and fish, page 71. Report No. C 78-18. Saskatchewan Research Council, Saskatoon, Saskatchewan. Surrounding water values are U238 = 50 ppb, Total Ra226 = 0.0009 pCi/g, Total Th = 1.7 ppb, Pb210 = 0.0026 pCi/g. To find a concentration in a plant or fish, multiply the surrounding water value times the value given.

[55] Robinson, D. J.; Ruggles, F. G.; Zaidi A. 1978. "A Study of Water Pollution in the Vicinity of Gulf Minerals Rabbit Lake, 1978." Environment Canada Surveillance Report EPS-5-W & NR-83-1 Western and Northern Region. About 100 pages. See pages i and 11. Available from: Environmental Protection Service, 241 Motherwell Building, 1901 Victoria Avenue, Regina, Saskatchewan, Canada. S4P 3R4.

[56] Ibid. See page i.

[57] Ibid. See page 11.

[58] Ibid. See pages 22, 31 and 32.

[59] Sheard, J.W. November, 1983. "Interim Report on Radionuclide Distribution in the Vegetation of Northern Saskatchewan to Saskatchewan Health Research Board." 26 pages. Department of Biology, University of Saskatchewan. A joint project

with the Saskatchewan Research Council.

[60] Op. Cit. Swanson, Stella M. September, 1982. See page 10.

[61] Op. Cit. Sheard, J. W. November, 1983. See pages 3 and 4.

[62] Aird, Paul. November 14, 1986. "The Food Chain Put In Peril From Fallout." *The Globe and Mail,* page A7. Paul Aird, Professor of Forest Policy, University of Toronto, Toronto, Ontario, Canada.

[63] Holtzman, R.B. 1968. "Ra226 and the Natural Airborne Nuclides Pb210 and Po210 in Arctic Biota." In: Synder, W. (ed.). "Radiation Protection." Pergamon Press, N.Y., N.Y. Pages 1,087-1,096.

[64] Kaarnen, P. and Miettinen, J.K. 1966. "Po210 and Pb210 in Environmental Samples in Finland." 6 pages. Department of Radiochemistry, University of Helsinki, Finland. Reprinted in: Abery and Hungate (eds.). 1966. "Radioecological Concentration Processes – Proceedings of an International Symposium Held in Stockholm April 25-29, 1966." Symposium Publications Division, Pergamon Press, N.Y., N.Y. Session D, pages 279-280.

[65] Op. Cit. Swanson, Stella M. September, 1982. See page 19.

[66] Op. Cit. AECB, Advisory Panel On Tailings. Sept., 1978. See page 3.

[67] Environmental Protection Service, Canadian Ministry of Environment, Pacific Region. March, 1980. "Statement of Evidence to the B.C. RCUM, Phase VI: Environmental Impact." Page 19. Report EPS 7-PR-79-1.

[68] IAEA, Vienna. April, 1986. "Environmental Migration of Radium and other Contaminants Present in Liquid and Solid Wastes from the Mining and Milling of Uranium, Final Report of a Co-ordinated Research Programme Sponsored by the IAEA 1981-1985." IAEA-Tecdoc-370. 39 pages.

[69] Op. Cit. AECB, Advisory Panel On Tailings. Sept., 1978. See page 4.

[70] B.C. RCUM. January 15, 1980. "Transcript of Proceedings." Vol. 56, pages 10,008-10,009.

[71] See the 1982 annual report on the Rabbit Lake mine submitted to the AECB.

[72] Op. Cit. AECB. Sept., 1978. See page 16.

[73] Dreeson, D.R. Feb., 1978. "Uranium Mill Tailings - Environmental Implications." LASL Mini-review. 4 pages. See page 1. LASL-77-37. Los Alamos Scientific Laboratory, University of California, Los Alamos, New Mexico, USA 87545.

[74] Op. Cit. Swanson, Stella M. September, 1982. See pages 7, 10, and 13.

[75] AECB publications noted can be ordered free of charge by

writing to AECB, Box 1046, Ottawa, Canada. K1P 5S9.

[76] AECB. March 31, 1978. "Letter from J. H. Jennekens, Director-General, Operations Directorate to R. N. Taylor the president of Gulf Minerals Canada Ltd. regarding Gulf Minerals Canada Ltd. Rabbit Lake Mine, Mill Facility Operating Licence AECB-MFOL-105-0." 3 pages. AECB file reference 22-G-38.

[77] National Research Council of Canada, Associate Committee on Scientific Criteria for Environmental Quality. December, 1973. "Lead in the Canadian Environment." 116 pages. See page 113. Available from: Publications, NRCC/CNRC, Ottawa, Canada, K1A 0R6. Publication No. BY73-7(ES). NRCC No. 13682.

[78] Op. Cit. Eldorado Nuclear Ltd. March 7, 1985. See page 37.

[79] Op. Cit. AECB. March 31, 1978.

[80] Saskmont Engineering Company Ltd. March, 1982. "Interim Report On The Investigation Of Radionuclide Levels In The Rabbit Lake Drainage System." Prepared for Gulf Minerals Canada Ltd. Saskmont Engineering, 3311 Circle Centre Mall, 8th Street East, Saskatoon, Saskatchewan, Canada. S7H 4K1. Available from the AECB.

[81] Gulf Minerals Canada Ltd. August 18, 1982. Letter to Mr. R. R. Sentis, Saskatchewan Environment, Prince Albert, Saskatchewan. The letter is filed in the AECB library.

[82] Gulf Minerals Canada Ltd. 1982. "Rabbit Lake Mine and Mill Annual Report to the AECB." See section 3.2.3.

[83] Op. Cit. Robinson, D. J.; Ruggles, F. G.; Zaidi A. 1978.

[84] Chamney, Larry G. May 28, 1985. "Summary of Water Quality and Fish Analysis In Wollaston Lake Near The Rabbit Lake/Collin's Bay B-Zone Facility." 12 pages. See page 4. AECB file 37-19-4-0.

[85] Saskatchewan Environment, Mines Pollution Control Branch. February, 1985. "Rabbit Lake - Collin's Bay Uranium Developments: Environmental Experience." 14 pages. See page 5. Available from Randy Sentis, Director, Mines Pollution Control Branch, Saskatchewan Environment, Box 3003, Prince Albert, Saskatchewan, Canada. Tel. 306-922-2224.

[86] Ibid.

[87] Eldorado Nuclear Limited. May, 1985. "Information: Eldor Mines, Eldorado Nuclear Limited." 3 pages. See page 2. For more information write to: Eldorado Nuclear Limited, 400-225 Albert Street, Ottawa, Ontario, Canada. K1P 6A9. Tel. 613-238-5222.

[88] Government of Saskatchewan. July 29, 1981. "Transcripts of Hearings on the Gulf Minerals Collins Bay B-zone Develop-

ment Held at La Ronge, Saskatchewan July 29, 1981, Vols. 4-5, Morning Session." See Vol. 5, page 150.

THE RESISTANCE

Quote by N. M. Ediger
[1] Ediger, N. M. July, 1978. "The Nuclear Controversy In Canada." In: "Uranium Supply And Demand, Proceedings of the Third International Symposium held by The Uranium Institute, London, July 12–14, 1978." Pages 229–234. See page 229. Mining Journal Books Ltd. ISBN: 0-900117-14-X.

[1] These two statements are reprinted from: Leis, Diana (ed.) (Interviews by Tony Dzeylion and translation by Mary Ann Kkailther) August, 1984. "Wollaston Lake People Speak Out Against Uranium Mining." 5 pages. See pages 2 and 3.
[2] Statement by Terry Daniels: Government of Saskatchewan. July 28, 1981. "Transcripts of Hearings on the Gulf Minerals Collins Bay B-zone Development Held at La Ronge, Saskatchewan July 28, 1981, Vol. 2" See page 101. Statement by Emil Hansen: See Vol. 4, page 45.
[3] The hoped for meeting with the government never took place. Premier Grant Devine refused the request.
[4] George Mercredi is a native person who was paid by the mining companies to do their public relations in northern Saskatchewan.
[5] A longer version of Gerry Paschen's Journal is printed in Canadians for Responsible Northern Development (CRND) Newsletter No. 25. 1985. CRND, 11911 University Ave., Edmonton, Alberta, Canada. T6G 1Z6. Tel. 403-436-4913.
[6] *The Northerner*. March 26, 1986. "North Gets Major Power Projects." Front page and page 6.
[7] This study was done by FSIN consultants. It takes a position in support of uranium mining. Anti-uranium mining sentiments are mentioned, but given a low profile and not elaborated on.
[8] Sol Sanderson gives the impression here that it is technically possible to stop the spread of contamination from uranium mining. However, there is no known solution at this time. See the "What About The Wastes?" section for more information.
[9] From April to June 1984 John Graham and Miles Goldstick conducted a 52 day, eight country European speaking tour. Sweden's largest newspaper, *Dagens Nyheter*, printed a front page story (May 21, 1984) that falsely states John Graham is a

310

"Chief." This was done despite John emphasizing in the interview that he was not a Chief, correcting a false rumor that originated from a sensation-hungry journalist. The accusation that Graham and Goldstick were lobbying against hunting and trapping is completely false.

[10] This statement was made during an interview in Pinehouse made by SAND on August 6, 1985.

INTERNATIONAL SOLIDARITY AUG. – SEPT. 1985

[1] Lillebror, Grensen 8, 0159 Oslo, Norway.

[2] Some literature based on information gathered on the SAND tour includes:

– Fredriksson, Inger. October 10, 1985. "Reste fran Uddevalla Till Kanadas Ödemarker." ("Travelled from Uddevalla to Canada's wilderness.") *Udevalla Posten.* Nr. 41. Front page and page 10.

– Hertzberg, L. October 17, 1985. "Indianreservat i Kanada, Ont om indiankultur, gott om gruvbolag." ("Indian reserves in Canada, Bad on Indian culture, good on mining companies.") *Älvborgs-posten.* Page 19.

– Klas Göran Sannerman. August 27, 1985. "Birgitta Ohlsson om uranbrytningen i Kanada: 'En cynisk exploatering'." ("Birgitta Ohlsson on uranium mining in Canada: 'cynical exploitation'.") *Ljusnan.* Nr. 198. Front page and page 4.

– Lundqvist, Kia. 1986. "Finlands uranimport." ("Finland's importation of uranium.") Pages 96–102 in: Lindblad, Kjell, "Efter Tjernobyl." ("After Chernobyl.") 176 pages. Boklaget, Box 258, (SF-00131, Helsinki, Finland. ISBN: 951-828-666-3.

– Kovalainen, Ritva. 1986. "Maa Pitää Kaikki – Kaikki Pitää Maa." ("The Earth Is Everyone's – Everyone Is The Earth.") Limited edition photo portfolio with text in Finnish; 300 copies.

Index Of People Quoted And Mentioned

(The list is not comprehensive.)

313

Abbreviations And Glossary Of Terms

AECB – Atomic Energy Control Board; the Canadian government agency responsible for regulating the nuclear industry.

B.C. RCUM – British Columbia Royal Commission of Inquiry into Uranium Mining.

Bq – Becquerel; one radioactive disintegration per second.

Bq/g – Becquerel per gram.

Ci – Curie; one curie is the number of transformations or disintegrations of one gram of radium. One curie equals 3.7×10^{10} disintegrations per second.

DIAND – Department of Indian Affairs and Northern Development; the federal Canadian Ministry dealing with Indian people.

DU – Depleted uranium; uranium with almost all the uranium-235 removed, leaving almost pure uranium-238.

ENL – Eldorado Nuclear Limited; a Canadian government owned corporation which mines and refines uranium.

FSIN – Federation of Saskatchewan Indian Nations; the administrative organization for the Indian Bands in Saskatchewan recognized by the federal government.

isotope – An atom of a given element that has slightly different physical and chemical properties from other atoms of the same element.

KLMC – Key Lake Mining Corporation.

LAC – Local Advisory Council; the administrative body for villages in northern Saskatchewan.

pCi/g – Picocuries per gram, or 1×10^{-12} curies per gram.

314

pCi/l – Picocuries per litre, or 1×10^{-12} curies per litre.

ppb – Parts per billion.

RCMP – Royal Canadian Mounted Police; the federal Canadian police force.

SAND – Scandinavians Against Nuclear Development; a group of 11 Swedish, Norwegian, and Finnish people who made a three week tour of northern Saskatchewan in July and August, 1985.

SANLG – Saskatchewan Association of Northern Local Governments; an organization representing 22 of the north's 30 Metis and non-status Indian settlements.

SINCO – Saskatchewan Indian Nations' Corporation; a business enterprise of FSIN.

SMDC – Saskatchewan Mining Development Corporation; a resource exploitation company owned by the government of Saskatchewan.

yellowcake – The common name for ammonium diuranate or uranium oxide (U_3O_8), the marketable product from a uranium mill. It is a fine, sand-like, yellow material.

Write for free catalogue of more than 110 books:

Black Rose Books
3981, boul. St. Laurent
Montréal, Québec
H2W 1Y5

Printed by
the workers of
Imprimerie Gagné
Louiseville, Québec
for
Black Rose Books